SpringerBriefs in Applied Sciences and Technology

Continuum Mechanics

Series Editors

Holm Altenbach, Institut für Mechanik, Lehrstuhl für Technische Mechanik, Otto von Guericke University Magdeburg, Magdeburg, Germany

Andreas Öchsner, Faculty of Mechanical Engineering, Esslingen University of Applied Sciences, Esslingen am Neckar, Australia

These SpringerBriefs publish concise summaries of cutting-edge research and practical applications on any subject of Continuum Mechanics and Generalized Continua, including the theory of elasticity, heat conduction, thermodynamics, electromagnetic continua, as well as applied mathematics.

SpringerBriefs in Continuum Mechanics are devoted to the publication of fundamentals and applications, presenting concise summaries of cutting-edge research and practical applications across a wide spectrum of fields. Featuring compact volumes of 50 to 125 pages, the series covers a range of content from professional to academic.

Sergey Korobeynikov · Alexey Larichkin ·
Patrizio Neff

Two Types of Compressible Isotropic Neo-Hookean Material Models

Sergey Korobeynikov
Lavrentyev Institute of Hydrodynamics
Novosibirsk, Russia

Alexey Larichkin
Lavrentyev Institute of Hydrodynamics
Novosibirsk, Russia

Patrizio Neff
Faculty of Mathematics
University of Duisburg-Essen
Essen, Germany

ISSN 2191-530X   ISSN 2191-5318 (electronic)
SpringerBriefs in Applied Sciences and Technology
ISSN 2625-1329   ISSN 2625-1337 (electronic)
SpringerBriefs in Continuum Mechanics
ISBN 978-3-032-06049-5   ISBN 978-3-032-06050-1 (eBook)
https://doi.org/10.1007/978-3-032-06050-1

© The Editor(s) (if applicable) and The Author(s), under exclusive license to Springer Nature Switzerland AG 2026

This work is subject to copyright. All rights are solely and exclusively licensed by the Publisher, whether the whole or part of the material is concerned, specifically the rights of translation, reprinting, reuse of illustrations, recitation, broadcasting, reproduction on microfilms or in any other physical way, and transmission or information storage and retrieval, electronic adaptation, computer software, or by similar or dissimilar methodology now known or hereafter developed.
The use of general descriptive names, registered names, trademarks, service marks, etc. in this publication does not imply, even in the absence of a specific statement, that such names are exempt from the relevant protective laws and regulations and therefore free for general use.
The publisher, the authors and the editors are safe to assume that the advice and information in this book are believed to be true and accurate at the date of publication. Neither the publisher nor the authors or the editors give a warranty, expressed or implied, with respect to the material contained herein or for any errors or omissions that may have been made. The publisher remains neutral with regard to jurisdictional claims in published maps and institutional affiliations.

This Springer imprint is published by the registered company Springer Nature Switzerland AG
The registered company address is: Gewerbestrasse 11, 6330 Cham, Switzerland

If disposing of this product, please recycle the paper.

Preface

This book provides readers with a deep understanding of the principles for generating formulations of compressible isotropic hyperelastic material models based on formulations of incompressible material models. The reference high-performance incompressible isotropic hyperelastic material model is Ogden's model, for which the elastic energy is generally represented as the sum of elemental energies based on strain tensors from the Doyle–Ericksen family. For the sake of transparency, the study is confined to considering the elastic energy only for one term of this sum based on the Finger strain tensor corresponding to the well-known Neo-Hookean material model. The book presents a systematic study of the performance of two known types of compressible generalization of the incompressible Neo-Hookean material model. The first type of generalization is based on the development of volumetric-isochoric Neo-Hookean models and involves the additive decomposition of the elastic energy into volumetric and isochoric parts. The second, simpler type of generalization is based on the development of mixed Neo-Hookean models that do not use this decomposition. Theoretical studies of model performance and simulations of some homogeneous deformations have shown that when using "good" volumetric functions, mixed and volumetric-isochoric models show similar performance in applications and have physically reasonable responses in extreme states, which is convenient for theoretical studies. However, compared to volumetric-isochoric models, mixed models allow the use of a wider set of volumetric functions with physically reasonable responses in extreme states. Another feature of mixed models is that they allow for simpler expressions for stresses and tangent stiffness tensors. This book will be useful both for novice researchers in developing hyperelastic equations for compressible materials and for experienced researchers by providing a brief overview of methods for generating compressible hyperelastic formulations based on available incompressible hyperelastic formulations. The book will also be useful for developers of computer codes for implementing hyperelastic models in FE systems. In addition, this book will be of interest to users of commercial FE codes, since these codes are often so-called black boxes and this book shows how to test hyperelastic models for any sample under uniform deformation.

Novosibirsk, Russia Sergey Korobeynikov
Novosibirsk, Russia Alexey Larichkin
Essen, Germany Patrizio Neff
June 2025

Acknowledgements The authors thank Prof. Alexey V. Shutov (Lavrentyev Institute of Hydrodynamics, Novosibirsk, Russia) for his interest and fruitful discussions during the presentation of the content of this book in the seminar "Mechanics of macro- and nano-structures" at Lavrentyev Institute of Hydrodynamics and on-line in May'2025. The authors also thank M.Sc. Nina Husemann (Faculty of Mathematics, University of Duisburg-Essen, Essen, Germany) for some technical help in writing the text of this book.

Contents

1 **Introduction** 1
  1.1 State of the Art 1
  1.2 Goals of the Book 3
  References 6

2 **Preliminaries** 9
  2.1 Basic Equations of Tensor Algebra 9
  2.2 Local Body Deformations and Basic Kinematics 14
  2.3 Stress Tensors and Their Rates 16
  2.4 Elastic Energy and Constitutive Relations for the Linear
      Isotropic Elastic Model 17
      2.4.1 Constitutive Relations for Incompressible Materials ... 18
      2.4.2 Constitutive Relations for Compressible Materials:
            Coupled Representation of the Elastic Energy 18
      2.4.3 Constitutive Relations for Compressible Materials:
            Decoupled Representation of the Elastic Energy 19
  References 20

3 **Constitutive Relations for Neo-Hookean Isotropic Hyperelastic
  Material Models** 23
  3.1 Constitutive Relations for the Incompressible Neo-Hookean
      Isotropic Hyperelastic Material Model 23
  3.2 Constitutive Relations for the Compressible Mixed Isotropic
      Hyperelastic Neo-Hookean Material Models 25
  3.3 Constitutive Relations for the Compressible vol-iso Isotropic
      Hyperelastic Neo-Hookean Material Models 26
  3.4 Discussion of Expressions for Compressible Neo-Hookean
      Material Models 27
  References 30

4 Some Volumetric Functions and Their Properties ... 33
4.1 Properties of Volumetric Functions ... 33
4.2 Some Volumetric Functions ... 34
References ... 38

5 Constitutive Inequalities for Neo-Hookean Materials ... 41
5.1 Hill and CSP Constitutive Inequalities ... 41
5.2 Testing Linear Elastic Isotropic Material Models Using Drucker's Postulate—Convexity of the Energy ... 43
5.3 Testing Neo-Hookean Models Using the Hill Postulate ... 44
    5.3.1 Testing the Incompressible Neo-Hookean Material Model ... 44
    5.3.2 Testing the Compressible Mixed Material Models ... 46
    5.3.3 Testing the Compressible vol-iso Material Models ... 47
5.4 Testing Neo-Hookean Models Using the CSP ... 51
References ... 53

6 Testing Neo-Hookean Materials in Homogeneous Deformations ... 55
6.1 Forms of Stress and Strain/Deformation Tensors for Homogeneous Deformations ... 55
6.2 Uniaxial Loading ... 58
    6.2.1 Incompressible Isotropic Neo-Hookean Material ... 59
    6.2.2 Compressible Isotropic Mixed Neo-Hookean Material Models ... 59
    6.2.3 Compressible Isotropic vol-iso Neo-Hookean Material Models ... 64
6.3 Equibiaxial Loading in Plane Stress ... 67
    6.3.1 Incompressible Isotropic Neo-Hookean Material ... 68
    6.3.2 Compressible Isotropic Mixed Neo-Hookean Material Models ... 70
    6.3.3 Compressible Isotropic vol-iso Neo-Hookean Material Models ... 74
6.4 Uniaxial Loading in Plane Strain ... 76
    6.4.1 Incompressible Isotropic Neo-Hookean Material ... 77
    6.4.2 Compressible Isotropic Mixed Neo-Hookean Material Models ... 80
    6.4.3 Compressible Isotropic vol-iso Neo-Hookean Material Models ... 81
References ... 92

7	**Concluding Remarks**	93
	7.1 Comparative Analysis of the Performance of Mixed and vol-iso Material Models	93
	7.2 General Conclusions	96
	References	97
Index		99

Acronyms

CSP	Corotational stability postulate
FE	Finite element
TSTS-M$^+$	True Stress True Strain Hilbert-Monotonicity condition
1st PK	First Piola–Kirchhoff

Chapter 1
Introduction

Abstract The constitutive relations of *hyperelasticity* (or *Green elasticity*) have been developed for over 90 years; therefore, the state of the art of these studies is presented in Sect. 1.1. In Sect. 1.2 we formulate the objectives of this study and give a brief overview of the development of the considered two types of compressible isotropic neo-Hookean material models.

1.1 State of the Art

In large strain solid mechanics, by the constitutive relations of *hyperelasticity* (or *Green elasticity*) are meant the functional relations between stress and strain tensors using a scalar tensor function called the *elastic energy*, see, e.g., [1]. Currently, two approaches to generating elastic energies for isotropic hyperelastic material models exist and are being developed (see, e.g., [1–9]). The first and historically earlier approach is based on the use of invariants of the left (or right) Cauchy–Green deformation tensor in expressions for elastic energies (note here the outstanding contributions of Mooney and Rivlin, see, e.g., [1]). The second approach is based on the use of principal stretches as independent variables for elastic energies (the first model to use this approach is apparently the Hencky one [10, 11]).

Because of the ease of mathematical derivation of elastic energy expressions and due to experimental studies on the deformation of nearly incompressible rubber-like materials, early hyperelastic models were based on the simplifying assumption of incompressibility of the materials under study. However, the subsequent development of the theoretical foundations of hyperelasticity, driven primarily by numerous applications to large deformations of compressible materials (e.g., elastic foams, graphene, biological tissues, etc.), finite element (FE) implementations of hyperelasticity models,[1] and the realization that purely incompressible materials do not exist in nature, led to the study of the properties of compressible (or slightly

---

[1] In FE-implementations, constitutive relations for incompressible materials are typically replaced by constitutive relations for compressible ones, this replacement is a consequence of using the penalty function method to satisfy the incompressibility condition (see, e.g., [3–5, 12]).

© The Author(s), under exclusive license to Springer Nature Switzerland AG 2026
S. Korobeynikov et al., *Two Types of Compressible Isotropic Neo-Hookean Material Models*, SpringerBriefs in Continuum Mechanics,
https://doi.org/10.1007/978-3-032-06050-1_1

compressible) hyperelastic material models. Naturally, researchers began to formulate elastic energy expressions for compressible materials based on the well-known elastic energy expressions for incompressible materials.

Within the framework of linear elastic theory, the elastic energy for compressible isotropic materials can be represented in both decoupled and coupled theoretically equivalent forms. In the first case, the elastic energy has an additive representation as a sum of purely volumetric and purely isochoric energies, and in the second case, the same elastic energy can be represented as a sum of purely volumetric and mixed—volumetric and isochoric—energies. Therefore, when generalizing incompressible hyperelastic material models to account for material compressibility, one faces a dilemma: how to represent the elastic energy: in the decoupled or coupled form? This dilemma arises, e.g., when the Ogden model [13], originally developed for incompressible materials, is generalized to account for material compressibility.[2] In the original generalization of his model, Ogden used the coupled form of elastic energy [14] (see also [8, 12]), and in the subsequent generalization, Ogden used the uncoupled form of elastic energy [7] (see also [4–6]). Note that, unlike the linear elastic model, the generalization of the elastic energy for the hyperelastic incompressible material model to account for material compressibility in the decoupled and coupled forms leads to two different material models. We observe that the generalization of the elastic energy in the coupled form seems to be simpler and more elegant than the generalization in the decoupled form, since in this case, the term containing the Lagrange multiplier in the potential energy expression for incompressible material is simply replaced by the volumetric energy, whereas in the second case, one uses the complex technique of multiplicative decomposition of principal stretches in volumetric-isochoric form proposed by Richter in the late 1940s–early 1950s (cf., [15]).[3] Compressible hyperelastic material models with elastic energies in the coupled form are studied in [8, 14, 22–35], and models of the same material with elastic energies in the decoupled form are addressed in [1, 3, 5, 6, 8, 9, 24, 27–32, 34, 36–40]. We emphasize that both formulations are rank-one convex and polyconvex provided that μ, K, $\lambda > 0$ and the volumetric function $h(J)$ is convex in $J = \det \mathbf{F}$ (cf., [41]), where μ is the shear modulus, K is the bulk modulus, λ is the second Lamé parameter, and $\mathbf{F}$ is the deformation gradient.

There are also other approaches to the compressible generalization of elastic energies for hyperelastic incompressible materials, although based on the Richter–Flory decomposition, but not involving the additive decomposition of elastic energies into volumetric and isochoric strain energies (see, e.g., [7, 22, 23, 42–46]). Such elastic energies are used to capture the features of dilatation behavior of slightly compressible rubber-like materials during deformation, in particular, to obtain agreement with the experimental findings by Penn [39], who established that the use of decoupled

---

[2] The Ogden model uses principal stretches as independent variables.

[3] Note that this decomposition technique is often attributed to Flory [16], whose work was published more than 10 years after the publications by Richter [17–20] in Germany (cf., [15], see also [21]).

forms of elastic energies is inconsistent with his experimental data on the dependence of dilatation on elongation under uniaxial loading. However, the analysis of this type of elastic energies is beyond the scope of this book.

1.2 Goals of the Book

Since the decoupled form of elastic energy expressions contradicts some experimental data, the question arises: do hyperelasticity models based on the decoupled form of elastic energy expressions indeed have such a decisive advantage over similar models based on the direct generalization of elastic energy expressions without using the Richter–Flory decomposition, which led to the fact that models of the first type are implemented in many commercial FE systems to the detriment of models of the second type for simulating deformations of slightly compressible rubber-like materials? The answer to this question can be obtained by comparing the performance characteristics of the above two types of hyperelastic material models, especially under conditions of slight compressibility. It is reasonable to restrict this comparison to compressible generalizations of elastic energy expressions for the simplest incompressible isotropic hyperelastic material model—the neo-Hookean material model. First, the elastic energy for this model is written as a dependence on the first invariant of the right (or left) Cauchy–Green deformation tensor (see, e.g., [1]), and, second, this elastic energy can be represented as a one-power version of the Ogden material with power $n = 2$ (cf., [13]). We consider therefore two families of elastic energies for compressible neo-Hookean material models, the mixed models and the vol-iso ones. Both are then composed with the same volumetric functions $h(J)$ (see Fig. 1.1). For each considered $h(J)$ we give a number #1,...,#8. In this way, speaking of a mixed or vol-iso model with #k ($k = 1, \ldots, 8$), the energy is uniquely defined (see Tables 1.1 and 1.2), where $J = \det \mathbf{F}$.

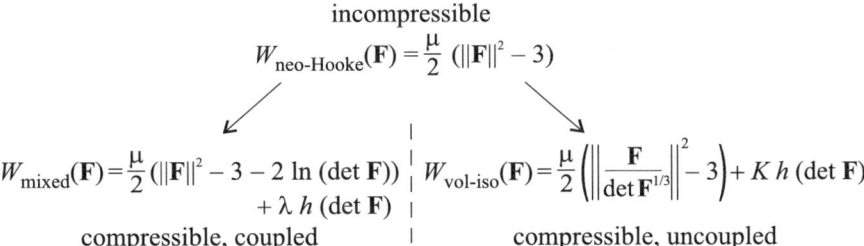

Fig. 1.1 Sketch of generation of two different types of the elastic energies W_{mixed} and $W_{\text{vol-iso}}$ (**F** is the deformation gradient) for compressible neo-Hookean material models based on the elastic energy $W_{\text{neo-Hooke}}$ for the incompressible neo-Hookean material model (see Sect. 3 for details). Here, μ is the standard shear modulus, λ is the second Lamé parameter and K is the bulk modulus

Table 1.1 Some elastic energies for the mixed compressible neo-Hookean materials

Model ID	Expression for the elastic energy
1	$\mu\,(\|\mathbf{F}\|^2 - 3 - 2\ln J)/2 + \lambda\,(\ln J)^2/2$
2	$\mu\,(\|\mathbf{F}\|^2 - 3 - 2\ln J)/2 + \lambda\,(J + J^{-1} - 2)/2$
3	$\mu\,(\|\mathbf{F}\|^2 - 3 - 2\ln J)/2 + \lambda\,(J^2 + J^{-2} - 2)/8$
4	$\mu\,(\|\mathbf{F}\|^2 - 3 - 2\ln J)/2 + \lambda\,(J^5 + J^{-5} - 2)/50$
5	$\mu\,(\|\mathbf{F}\|^2 - 3 - 2\ln J)/2 + \lambda\,(J^2 - 2\ln J - 1)/4$
6	$\mu\,(\|\mathbf{F}\|^2 - 3 - 2\ln J)/2 + \lambda\,(J - \ln J - 1)$
7	$\mu\,(\|\mathbf{F}\|^2 - 3 - 2\ln J)/2 + \lambda\,(J - 1)^2/2$
8	$\mu\,(\|\mathbf{F}\|^2 - 3 - 2\ln J)/2 + \lambda\,(e^{\ln^2 J} - 1)/2$

Table 1.2 Some elastic energies for the vol-iso compressible neo-Hookean materials

Model ID	Expression for the elastic energy
1	$\mu\,(\|\mathbf{F}/J^{1/3}\|^2 - 3)/2 + K\,(\ln J)^2/2$
2	$\mu\,(\|\mathbf{F}/J^{1/3}\|^2 - 3)/2 + K\,(J + J^{-1} - 2)/2$
3	$\mu\,(\|\mathbf{F}/J^{1/3}\|^2 - 3)/2 + K\,(J^2 + J^{-2} - 2)/8$
4	$\mu\,(\|\mathbf{F}/J^{1/3}\|^2 - 3)/2 + K\,(J^5 + J^{-5} - 2)/50$
5	$\mu\,(\|\mathbf{F}/J^{1/3}\|^2 - 3)/2 + K\,(J^2 - 2\ln J - 1)/4$
6	$\mu\,(\|\mathbf{F}/J^{1/3}\|^2 - 3)/2 + K\,(J - \ln J - 1)$
7	$\mu\,(\|\mathbf{F}/J^{1/3}\|^2 - 3)/2 + K\,(J - 1)^2/2$
8	$\mu\,(\|\mathbf{F}/J^{1/3}\|^2 - 3)/2 + K\,(e^{\ln^2 J} - 1)/2$

At a sufficiently high degree of material compressibility, for example, at Poisson's ratio $\nu = 0.125$ (see Sect. 3.4), the two types of material models considered give different plots of elastic energies versus longitudinal stretches (see Fig. 1.2) when using the same volumetric functions in the uniaxial loading problem and the simple vol-iso model #7 even leads to an unphysical non-monotonic plot of the elastic energy versus longitudinal stretch (see curve #3 in Fig. 1.2c). This suggests that for the same two values of material parameters (shear modulus μ and Poisson's ratio ν) and the same volumetric function $h(J)$, these two types of material models can lead to different quantitative values of stresses in deformable bodies (see Sect. 6 for examples). This fact motivates a thorough and comprehensive study of the performance of the two types of compressible neo-Hookean material models under consideration.

One objective of this study is a comparative analysis of the performance of two types (*vol-iso/mixed*) of compressible isotropic hyperelastic neo-Hookean material models obtained by generalizing the standard neo-Hookean isotropic incompressible material model to account for volumetric energy with/without the additive decomposition of a elastic energy into volumetric and isochoric strain energies. In this comparison, we will check whether both types of compressible material models satisfy two fundamental postulates of solid mechanics formulated to test nonlinear material models for satisfaction of some desired properties not following from thermody-

1.2 Goals of the Book

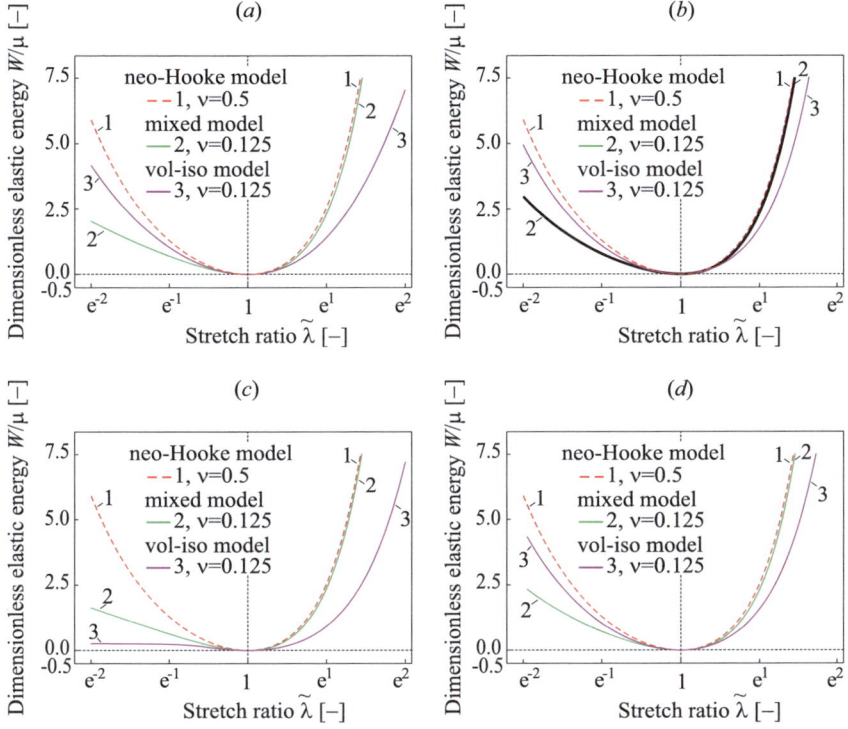

Fig. 1.2 Plots of elastic energies versus longitudinal stretches $\tilde{\lambda}$ in the uniaxial loading problem for compressible neo-Hookean models (a dimensionless elastic energy W/μ depends on Poisson's ratio ν only due to the expressions $\lambda = 2\mu\nu/(1-2\nu)$ ($\Leftrightarrow$ $\nu = \lambda/2(\lambda + \mu)$) and $K = \lambda + 2\mu/3$) with the following volumetric functions ($J = \det \mathbf{F}$) $h(J)$: $(\ln^2 J)/2$ **a**, $(J^5 + J^{-5})/50$ **b**, $(J-1)^2/2$ **c**, and $(e^{\ln^2 J} - 1)/2$ **d** (corresponding to volumetric functions #1, #4, #7, and #8, see Table 4.1 for the volumetric functions definition)

namic constraints. The first postulate (the *Hill postulate*), requiring the satisfaction of some inequality, was proposed more than half a century ago by Hill (cf., [47–49]), and the second inequality in the form of the corotational stability postulate (CSP) was proposed recently by Neff et al. (cf., [21, 50, 51]). Both these postulates extend the *Drucker postulate* (the material stability postulate), well-known in solid mechanics and usually formulated for infinitesimal strains, to the case of finite strains. The satisfaction of the first (Hill) postulate is equivalent to that the Kirchhoff stress tensor is a monotonic function of the logarithmic strain tensor, and the satisfaction of the second postulate (CSP) is equivalent to that the Cauchy stress tensor is a monotonic function of the logarithmic strain tensor.

The next objective of this study is to compare the performance of the above two types of compressible material models by solving some problems with homogeneous deformations. Note that the performance characteristics of these models in solving similar problems have already been studied by a number of researchers (see, e.g., [27, 29–31, 34, 38]), but the comparative analysis of the performance characteristics in these studies is apparently not thorough enough and not always fully correct. For example, a comparison of stresses and strains was performed [27, 31] using different types of volumetric energies in different models for the same homogeneous deformations, but for a correct comparison, it is necessary to use the same types of this energy for different models. In addition, previous studies have found some physically unreasonable responses for both types of material models under consideration. Therefore, our task here is to select the types of volumetric energy that would minimize these physically unreasonable responses.

In summary, the main objectives of this study are to answer the following questions:

1. what types of volumetric energies minimize the effects of physically unreasonable response for the neo-Hookean models under consideration?
2. is the mixed neo-Hookean compressible hyperelastic material model taking into account volumetric energy without the additive decomposition of the elastic energy into volumetric and isochoric strain energies inferior in performance to the vol-iso counterpart with this decomposition?

The novelty of this study lies in conclusively answering these questions.

References

1. Hackett, R.M.: Hyperelasticity Primer, 2nd edn. Springer, Cham (2018)
2. Bertram, A.: Elasticity and Plasticity of Large Deformations, 4th edn. Springer, Cham (2021)
3. Crisfield, M.A.: Non-linear Finite Element Analysis of Solids and Structures. In: Advanced Topics, vol. 2. Wiley, Chichester (1997)
4. de Borst, R., Crisfield, M.A., Remmers, J.J.C., Verhoosel, C.V.: Non-linear Finite Element Analysis of Solids and Structures, 2nd edn. Wiley, Chichester (2012)
5. de Souza Neto, E.A., Peric, D., Owen, D.J.R.: Computational Methods for Plasticity: Theory and Applications. Wiley, Chichester (2008)
6. Holzapfel, G.A.: Nonlinear Solid Mechanics: A Continuum Approach for Engineering. Wiley, Chichester (2000)
7. Ogden, R.W.: Non-linear Elastic Deformations. Ellis Horwood, Chichester (1984)
8. Wriggers, P.: Nonlinear Finite Element Methods. Springer, Berlin, Heidelberg (2008)
9. Zhang, J., Song, Y., Lu, B.: Mechanics of Elastic Solids. Springer, China (2025)
10. Hencky, H.: The elastic behavior of vulcanized rubber. Rubber Chem. Technol. **6**(2), 217–224 (1933). https://doi.org/10.5254/1.3547545
11. Hencky, H.: The elastic behaviour of vulcanized rubber. J. Appl. Mech. **1**(2), 45–53 (1933). https://doi.org/10.1115/1.4012174
12. Peng, S.H., Chang, W.V.: A compressible approach in finite element analysis of rubber-elastic materials. Comput. Struct. **62**(3), 573–593 (1997). https://doi.org/10.1016/S0045-7949(96)00195-2
13. Ogden, R.W.: Large deformation isotropic elasticity — on the correlation of theory and experiment for incompressible rubberlike solids. Proceedings of the Royal Society of London. A. Math. Phys. Sci. **326**(1567), 565–584 (1972). https://doi.org/10.1098/rspa.1972.0026

References

14. Ogden, R.W.: Large deformation isotropic elasticity: on the correlation of theory and experiment for compressible rubberlike solids. Proceedings of the Royal Society of London. A. Math. Phys. Sci. **328**(1575), 567–583 (1972). https://doi.org/10.1098/rspa.1972.0096
15. Graban, K., Schweickert, E., Martin, R.J., Neff, P.: A commented translation of Hans Richter's early work "The isotropic law of elasticity." Math. Mech. Solids **24**(8), 2649–2660 (2019). https://doi.org/10.1177/1081286519847495
16. Flory, P.J.: Thermodynamic relations for high elastic materials. Trans. Faraday Soc. **57**, 829–838 (1961). https://doi.org/10.1039/TF9615700829
17. Richter, H.: Das isotrope Elastizitätsgesetz. Z. Angew. Math. Mech. **28**(7–8), 205–209 (1948). https://doi.org/10.1002/zamm.19480280703
18. Richter, H.: Hauptaufsätze: Verzerrungstensor, Verzerrungsdeviator und Spannungstensor bei endlichen Formänderungen. Z. Angew. Math. Mech. **29**(3), 65–75 (1949). https://doi.org/10.1002/zamm.19490290301
19. Richter, H.: Zum Logarithmus einer Matrix. Arch. Math. **2**(5), 360–363 (1949). https://doi.org/10.1007/BF02036865
20. Richter, H.: Zur Elastizitätstheorie endlicher Verformungen. Mathematische Nachrichten **8**(1), 65–73 (1952). https://doi.org/10.1002/mana.19520080109
21. Neff, P., Holthausen, S., d'Agostino, M.V., Bernardini, D., Sky, A., Ghiba, I.D., Martin, R.J.: Hypo-elasticity, Cauchy-elasticity, corotational stability and monotonicity in the logarithmic strain. J. Mech. Phys. Solids **202**, 106074 (2025). https://doi.org/10.1016/j.jmps.2025.106074
22. Attard, M.M.: Finite strain-isotropic hyperelasticity. Int. J. Solids Struct. **40**(17), 4353–4378 (2003). https://doi.org/10.1016/S0020-7683(03)00217-8
23. Attard, M.M., Hunt, G.W.: Hyperelastic constitutive modeling under finite strain. Int. J. Solids Struct. **41**(18), 5327–5350 (2004). https://doi.org/10.1016/j.ijsolstr.2004.03.016
24. Bonet, J., Wood, R.D.: Nonlinear Continuum Mechanics for Finite Element Analysis, 2nd edn. Cambridge University Press, Cambridge (2008)
25. Chaves, E.W.: Notes on Continuum Mechanics. Springer, Barcelona (2013)
26. Clayton, J., Bliss, K.: Analysis of intrinsic stability criteria for isotropic third-order Green elastic and compressible neo-Hookean solids. Mech. Mater. **68**, 104–119 (2014). https://doi.org/10.1016/j.mechmat.2013.08.007
27. Ehlers, W., Eipper, G.: The simple tension problem at large volumetric strains computed from finite hyperelastic material laws. Acta Mech. **130**(1), 17–27 (1998). https://doi.org/10.1007/BF01187040
28. Hashiguchi, K., Yamakawa, Y.: Introduction to Finite Strain Theory for Continuum Elasto-Plasticity. Wiley, Hoboken (2013)
29. Kellermann, D.C., Attard, M.M.: An invariant-free formulation of neo-Hookean hyperelasticity. Z. Angew. Math. Mech. **96**(2), 233–252 (2016). https://doi.org/10.1002/zamm.201400210
30. Korobeynikov, S.N.: Families of Hooke-like isotropic hyperelastic material models and their rate formulations. Arch. Appl. Mech. **93**(10), 3863–3893 (2023). https://doi.org/10.1007/s00419-023-02466-5
31. Pence, T.J., Gou, K.: On compressible versions of the incompressible neo-Hookean material. Math. Mech. Solids **20**(2), 157–182 (2015). https://doi.org/10.1177/1081286514544258
32. Simo, J.C., Hughes, T.J.R.: Computational Inelasticity. Springer, Berlin (1998)
33. Simo, J.C., Pister, K.S.: Remarks on rate constitutive equations for finite deformation problems: computational implications. Comput. Methods Appl. Mech. Eng. **46**(2), 201–215 (1984). https://doi.org/10.1016/0045-7825(84)90062-8
34. Springhetti, R., Rossetto, G., Bigoni, D.: Buckling of thin-walled cylinders from three dimensional nonlinear elasticity. J. Elast. **154**(1), 297–323 (2023). https://doi.org/10.1007/s10659-022-09905-4
35. Stickle, M.M., Molinos, M., Navas, P., Yagüe, A., Manzanal, D., Moussavi, S., Pastor, M.: A component-free Lagrangian finite element formulation for large strain elastodynamics. Comput. Mech. **69**(3), 639–660 (2022). https://doi.org/10.1007/s00466-021-02107-0
36. Destrade, M., Gilchrist, M.D., Motherway, J., Murphy, J.G.: Slight compressibility and sensitivity to changes in Poisson's ratio. Int. J. Numer. Meth. Eng. **90**(4), 403–411 (2012). https://doi.org/10.1002/nme.3326

37. Haupt, P.: Continuum Mechanics and Theory of Materials, 2nd edn. Springer, Berlin (2002)
38. Kossa, A., Valentine, M.T., McMeeking, R.M.: Analysis of the compressible, isotropic, neo-Hookean hyperelastic model. Meccanica **58**(1), 217–232 (2023). https://doi.org/10.1007/s11012-022-01633-2
39. Penn, R.W.: Volume changes accompanying the extension of rubber. Transactions of the Society of Rheology **14**, 509–517 (1970). https://doi.org/10.1122/1.549176
40. Rubin, M.B.: Continuum Mechanics with Eulerian Formulations of Constitutive Equations. Springer, Cham (2021)
41. Hartmann, S., Neff, P.: Polyconvexity of generalized polynomial-type hyperelastic strain energy functions for near-incompressibility. Int. J. Solids Struct. **40**(11), 2767–2791 (2003). https://doi.org/10.1016/S0020-7683(03)00086-6
42. Fong, J.T., Penn, R.W.: Construction of a strain-energy function for an isotropic elastic material. Trans. Soc. Rheol. **19**(1), 99–113 (1975). https://doi.org/10.1122/1.549389
43. Huang, Z.P.: A novel constitutive formulation for rubberlike materials in thermoelasticity. J. Appl. Mech. **81**(4), 041013 (2014). https://doi.org/10.1115/1.4025272
44. Huang, Z.P.: Erratum: "A novel constitutive formulation for rubberlike materials in thermoelasticity" [asme j. appl. mech., 2014, 81(4), p. 041013]. J. Appl. Mech. **83**(4), 047001 (2016). https://doi.org/10.1115/1.4032660
45. Rogovoy, A.: Effect of elastomer slight compressibility. Eur. J. Mech. A. Solids **20**(5), 757–775 (2001). https://doi.org/10.1016/S0997-7538(01)01170-6
46. Yao, Y., Chen, S., Huang, Z.: A generalized Ogden model for the compressibility of rubberlike solids. Philosophical Transactions of the Royal Society A: Mathematical, Physical and Engineering Sciences **380**(2234), 20210320 (2022). https://doi.org/10.1098/rsta.2021.0320
47. Hill, R.: On constitutive inequalities for simple materials–I. J. Mech. Phys. Solids **16**(4), 229–242 (1968). https://doi.org/10.1016/0022-5096(68)90031-8
48. Hill, R.: Constitutive inequalities for isotropic elastic solids under finite strain. Proceedings of the Royal Society of London. A. Math. Phys. Sci. **314**(1519), 457–472 (1970). https://doi.org/10.1098/rspa.1970.0018
49. Hill, R.: Aspects of invariance in solid mechanics. In: Yih, C.S. (ed.) Advances in Applied Mechanics, vol. 18, pp. 1–75. Academic Press, New York (1979). https://doi.org/10.1016/S0065-2156(08)70264-3
50. d'Agostino, M.V., Holthausen, S., Bernardini, D., Sky, A., Ghiba, I.D., Martin, R.J., Neff, P.: A constitutive condition for idealized isotropic Cauchy elasticity involving the logarithmic strain. J. Elast. **157**(1), 23 (2025). https://doi.org/10.1007/s10659-024-10097-2
51. Neff, P., Husemann, N.J., Nguetcho Tchakoutio, A.S., Korobeynikov, S.N., Martin, R.J.: The corotational stability postulate: Positive incremental Cauchy stress moduli for diagonal, homogeneous deformations in isotropic nonlinear elasticity. Int. J. Non-Linear Mech. **174**, 105033 (2025). https://doi.org/10.1016/j.ijnonlinmec.2025.105033

Chapter 2
Preliminaries

Abstract In this chapter, we present the basic expressions for tensor algebra (Sect. 2.1), local body deformations and basic kinematics (Sect. 2.2), and stress tensors and their rates (Sect. 2.3) required for the exposition of the basic material of the research. In addition, Sect. 2.4 presents different forms of the elastic energy and constitutive relations for the linear isotropic material model that inspired the derivation of similar forms for the nonlinear neo-Hookean material (see Chap. 3).

2.1 Basic Equations of Tensor Algebra

We define the second-order tensor $\mathbf{H} \in \mathcal{T}^2$ and the fourth-order tensor $\mathbb{H} \in \mathcal{T}^4$ (hereinafter, $\mathcal{T}^2$ and $\mathcal{T}^4$ denote the sets of all second-order and fourth-order tensors). Hereinafter, $\mathcal{T}^2_{\text{sym}}$, $\mathcal{T}^2_{\text{skew}} \subset \mathcal{T}^2$ denote the sets of all symmetric and skew-symmetric second-order tensors, respectively; $\text{sym}\,\mathbf{A} \equiv (\mathbf{A} + \mathbf{A}^T)/2 \in \mathcal{T}^2_{\text{sym}}$ and $\text{skew}\,\mathbf{A} \equiv (\mathbf{A} - \mathbf{A}^T)/2 \in \mathcal{T}^2_{\text{skew}}$ denote the symmetric and skew-symmetric components of the tensor $\mathbf{A} \in \mathcal{T}^2$, respectively. Next, the set of fourth-order tensors with major symmetry will be denoted by $\mathcal{T}^4_{\text{Sym}} \subset \mathcal{T}^4$, the set of the same tensors with twice minor symmetry will be denoted by $\mathcal{T}^4_{\text{sym}} \subset \mathcal{T}^4$, and the set of tensors with both major and twice minor symmetries[1] by $\mathcal{T}^4_{\text{Ssym}} \subset \mathcal{T}^4$.

Let $\mathbf{A}, \mathbf{H} \in \mathcal{T}^2$. We define the *double inner product* (*double contraction*) operation of tensors:

$$\mathbf{A} : \mathbf{H} = \mathbf{H} : \mathbf{A} \equiv \text{tr}(\mathbf{A} \cdot \mathbf{H}^T) = \text{tr}(\mathbf{A}^T \cdot \mathbf{H}) = \text{tr}(\mathbf{H} \cdot \mathbf{A}^T) = \text{tr}(\mathbf{H}^T \cdot \mathbf{A}) \,(= \langle \mathbf{A}, \mathbf{H} \rangle).$$

Hereinafter, the superscript T denotes the transposition of a tensor, and the dot between vectors and/or tensors denotes their inner (matrix) product. Using this definition, we can introduce the norm of any second order tensor

---

[1] Hereinafter, tensors with both major and twice minor symmetries will be called *supersymmetric* (cf., [1]) or *fully symmetric* (cf., [2, 3]) tensors.

$$\|\mathbf{A}\| \equiv (\mathbf{A} : \mathbf{A})^{1/2} \ (= \sqrt{\langle \mathbf{A}, \mathbf{A} \rangle}).$$

We define the following four external product operations for second-order tensors according to the definitions given in [4–6]: *dyadic* $\mathbf{A} \otimes \mathbf{H}$, *direct* $\mathbf{A} \underline{\otimes} \mathbf{H}$, *alternate* $\mathbf{A} \overline{\otimes} \mathbf{H}$, and *symmetric* $\mathbf{A} \overset{\text{sym}}{\otimes} \mathbf{H} \equiv \frac{1}{2}(\mathbf{A} \underline{\otimes} \mathbf{H} + \mathbf{A} \overline{\otimes} \mathbf{H})$ tensor products.

Let $\mathbf{A}, \mathbf{B}, \mathbf{X} \in \mathcal{T}^2$. We can show that the following identities hold:

$$(\mathbf{A} \otimes \mathbf{B}) : \mathbf{X} = \mathbf{A}(\mathbf{B} : \mathbf{X}),$$

$$(\mathbf{A} \overset{\text{sym}}{\otimes} \mathbf{B}) : \mathbf{X} = \mathbf{A} \cdot \text{sym} \, \mathbf{X} \cdot \mathbf{B}^T.$$

Let $\mathbf{S} \in \mathcal{T}^2_{\text{sym}}$. This tensor can be represented in the classical spectral form

$$\mathbf{S} = \sum_{k=1}^{3} s_k \, \mathbf{n}_k \otimes \mathbf{n}_k, \tag{2.1}$$

where $s_k \in \mathbb{R}$ are eigenvalues and $\{\mathbf{n}_k\}$ ($k = 1, 2, 3$) is the triad of the corresponding subordinate right-oriented orthonormal eigenvectors (corresponding to the principal directions) of the tensor $\mathbf{S}$, and the symbol $\otimes$ denotes the dyadic product of vectors. For multiple eigenvalues s_k, the corresponding eigenvectors $\mathbf{n}_k$ are defined ambiguously. This ambiguity can be circumvented by representing the tensor $\mathbf{S}$ in terms of *eigenprojections* (see, e.g., [7–10]):

$$\mathbf{S} = \sum_{i=1}^{m} s_i \, \mathbf{S}_i. \tag{2.2}$$

Here, s_i are all different m^2 eigenvalues of the tensor $\mathbf{S}$ and $\mathbf{S}_i$ ($i = 1, \ldots, m$) are the subordinate *eigenprojections*. Without loss of generality, we number the eigenvalues s_k and define the eigenprojections depending on the eigenindex m as follows:

$$m = \begin{cases} 3, & s_1 \neq s_2 \neq s_3 \neq s_1, \quad \mathbf{S}_i = \mathbf{n}_i \otimes \mathbf{n}_i \ (i = 1, 2, 3); \\ 2, & s_1 \neq s_2 = s_3, \quad \mathbf{S}_1 = \mathbf{n}_1 \otimes \mathbf{n}_1, \ \mathbf{S}_2 = \mathbf{n}_2 \otimes \mathbf{n}_2 + \mathbf{n}_3 \otimes \mathbf{n}_3 = \mathbf{I} - \mathbf{S}_1; \\ 1, & s_1 = s_2 = s_3, \quad \mathbf{S}_1 = \mathbf{n}_1 \otimes \mathbf{n}_1 + \mathbf{n}_2 \otimes \mathbf{n}_2 + \mathbf{n}_3 \otimes \mathbf{n}_3 = \mathbf{I}. \end{cases} \tag{2.3}$$

Hereinafter, $\mathbf{I}$ denotes the identity tensor. The eigenprojections have the following properties [10]:

$$\mathbf{S}_i \cdot \mathbf{S}_j = \begin{cases} \mathbf{S}_i & \text{if } i = j \\ \mathbf{0} & \text{if } i \neq j \end{cases}, \quad \sum_{i=1}^{m} \mathbf{S}_i = \mathbf{I}, \quad \text{tr} \, \mathbf{S}_i = m_i \quad (i, j = 1, \ldots, m). \tag{2.4}$$

---

[2] The number m ($1 \leq m \leq 3$) will be called the *eigenindex*.

2.1 Basic Equations of Tensor Algebra

Here, m_i denotes the multiplicity of an eigenvalue s_i and $\mathbf{0} \in \mathcal{T}^2$ denotes the zero second-order tensor.

Let the tensors $\mathbf{S}, \mathbf{H} \in \mathcal{T}^2_{\text{sym}}$ be arbitrary tensors, and let the tensor $\mathbf{S}$ have the spectral representation (2.2). We represent the tensor $\mathbf{H}$ in the following form (see Corollary 2.1 of [11]):[3]

$$\mathbf{H} = \hat{\mathbf{H}} + \tilde{\mathbf{H}}, \qquad \hat{\mathbf{H}} \equiv \sum_{i=1}^{m} \mathbf{S}_i \cdot \mathbf{H} \cdot \mathbf{S}_i, \qquad \tilde{\mathbf{H}} \equiv \sum_{i \neq j=1}^{m} \mathbf{S}_i \cdot \mathbf{H} \cdot \mathbf{S}_j, \qquad (2.5)$$

so that $\hat{\mathbf{H}}, \tilde{\mathbf{H}} \in \mathcal{T}^2_{\text{sym}}$ are the components of the tensor $\mathbf{H}$ that are coaxial and orthogonal[4] to the tensor $\mathbf{S}$. Since the tensor $\hat{\mathbf{H}}$ is coaxial with the tensor $\mathbf{S}$, this tensor can be also represented as

$$\hat{\mathbf{H}} = \sum_{i=1}^{m} H_i \mathbf{S}_i. \qquad (2.6)$$

Proposition 2.1 *Suppose that $\mathbf{H}, \mathbf{S} \in \mathcal{T}^2_{\text{sym}}$; $\mathbf{S}$ has the spectral representation (2.2); and the tensor $\mathbf{X}$ is an isotropic tensor function of the tensor arguments $\mathbf{S}$ and $\mathbf{H}$ which is linear in $\mathbf{H}$ and can be written as ($2 \leq m \leq 3$)*

$$\mathbf{X}(\mathbf{S}, \mathbf{H}) = \sum_{i \neq j=1}^{m} x_{ij} \, \mathbf{S}_i \cdot \mathbf{H} \cdot \mathbf{S}_j, \quad x_{ij} \equiv x(s_i, s_j) \; (x_{ji} = x_{ij}),$$

or, in an alternative form, as[5]

$$\mathbf{X}(\mathbf{S}, \mathbf{H}) = \mathbb{X}(\mathbf{S}) : \mathbf{H} \quad (\mathbb{X} \in \mathcal{T}^4_{\text{Ssym}}), \qquad (2.7)$$

where

$$\mathbb{X}(\mathbf{S}) \equiv \sum_{i \neq j=1}^{m} x_{ij} \, \mathbf{S}_i \overset{\text{sym}}{\otimes} \mathbf{S}_j.$$

Then the inequalities

$$x_{ij} > 0 \quad (i, j = 1, \ldots m, \; i \neq j)$$

are necessary and sufficient for the positive definiteness of the tensorial function $\mathbf{X}(\mathbf{S}, \mathbf{H})$ and its associated tensor $\mathbb{X}$ with respect to the tensor $\tilde{\mathbf{H}}$ (which is orthogonal to the tensor $\mathbf{S}$); i.e., $\mathbf{X} : \tilde{\mathbf{H}} > 0 \; \forall \; \tilde{\mathbf{H}} \neq \mathbf{0} \Leftrightarrow \tilde{\mathbf{H}} : \mathbb{X} : \tilde{\mathbf{H}} > 0 \; \forall \; \tilde{\mathbf{H}} \neq \mathbf{0}$.

---

[3] Hereinafter, the notation $\sum_{i \neq j=1}^{m}$ denotes the summation over $i, j = 1, \ldots, m$ and $i \neq j$ and this summation is assumed to vanish when $m = 1$.
[4] Tensors $\mathbf{X}, \mathbf{Y} \in \mathcal{T}^2_{\text{sym}}$ will be called orthogonal if the equality $\mathbf{X} : \mathbf{Y} = 0$ is satisfied.
[5] The fact that $\mathbb{X} \in \mathcal{T}^4_{\text{Ssym}}$ follows from the statements of Theorem 2.1 in [11].

Proof. In view of the statements of Theorem 2.2 in [11], the tensor $\mathbf{X}$ is also orthogonal to the tensor $\mathbf{S}$, whence it follows that

$$\mathbf{X} : \mathbf{H} = \mathbf{X} : \tilde{\mathbf{H}}.$$

Since the tensors considered are coaxial and orthogonal to the tensor $\mathbf{S}$ and hence to each other, we obtain the equality

$$\mathbf{H} : \mathbb{X} : \mathbf{H} = \tilde{\mathbf{H}} : \mathbb{X} : \tilde{\mathbf{H}}.$$

In view of the symmetry of the quantities $x_{ij} > 0$ $(i, j = 1, \ldots m,\ i \neq j)$, the symmetry of the tensor $\tilde{\mathbf{H}}$ (which is a consequence of the symmetry of the tensor $\mathbf{H}$), and the representation of the eigenprojections in (2.3), direct calculations yield

$$\mathbf{H} : \mathbb{X} : \mathbf{H} = \tilde{\mathbf{H}} : \mathbb{X} : \tilde{\mathbf{H}} = \begin{cases} 2(x_{12} H_{12}^2 + x_{13} H_{13}^2 + x_{23} H_{23}^2), & \text{if } m = 3 \\ 2 x_{12}(H_{12}^2 + H_{13}^2), & \text{if } m = 2 \end{cases}, \quad (2.8)$$

where

$$H_{kl}(= H_{lk}) \equiv \mathbf{n}_k \cdot \mathbf{H} \cdot \mathbf{n}_l \ (k, l = 1, 2, 3,\ k \neq l).$$

The statements of Proposition follow from (2.8). □

Example 2.1
Let us consider a simple example for 2D analysis illustrating the statements of Proposition 2.1. Let the tensor $\mathbf{S} \in \mathcal{T}_{\text{sym}}^2$ have the following spectral representation of the form (2.1)

$$\mathbf{S} = s_1 \mathbf{n}_1 \otimes \mathbf{n}_1 + s_1 \mathbf{n}_2 \otimes \mathbf{n}_2.$$

Let any tensor $\mathbf{H} \in \mathcal{T}_{\text{sym}}^2$ have the following representation in the principal axes of the tensor $\mathbf{S}$:

$$\mathbf{H} = H_{11} \mathbf{n}_1 \otimes \mathbf{n}_1 + H_{22} \mathbf{n}_2 \otimes \mathbf{n}_2 + H_{12}(\mathbf{n}_1 \otimes \mathbf{n}_2 + \mathbf{n}_2 \otimes \mathbf{n}_1),$$

or

$$\mathbf{H} = \hat{\mathbf{H}} + \tilde{\mathbf{H}}, \quad \hat{\mathbf{H}} \equiv H_{11} \mathbf{n}_1 \otimes \mathbf{n}_1 + H_{22} \mathbf{n}_2 \otimes \mathbf{n}_2, \quad \tilde{\mathbf{H}} \equiv H_{12}(\mathbf{n}_1 \otimes \mathbf{n}_2 + \mathbf{n}_2 \otimes \mathbf{n}_1).$$

We define the following isotropic tensor function of its arguments $\mathbf{X}(\mathbf{S}, \mathbf{H}) \in \mathcal{T}_{\text{sym}}^2$ which is linear in $\mathbf{H}$

$$\mathbf{X}(\mathbf{S}, \mathbf{H}) = H_{12}(s_1 + s_2)(\mathbf{n}_1 \otimes \mathbf{n}_2 + \mathbf{n}_2 \otimes \mathbf{n}_1).$$

2.1 Basic Equations of Tensor Algebra

Note that this tensor function can be represented in the form (2.7), where

$$\mathbb{X}(\mathbf{S}) \equiv (s_1 + s_2)(\mathbf{n}_1 \otimes \mathbf{n}_1) \overset{\text{sym}}{\otimes} (\mathbf{n}_2 \otimes \mathbf{n}_2).$$

Using the definition of the symmetric external tensor product operation [4], we obtain an explicit representation of the supersymmetric tensor $\mathbb{X}(\mathbf{S})$ in the principal axes of the tensor $\mathbf{S}$

$$\mathbb{X}(\mathbf{S}) = \frac{1}{2}(s_1 + s_2)(\mathbf{n}_1 \otimes \mathbf{n}_2 \otimes \mathbf{n}_1 \otimes \mathbf{n}_2 + \mathbf{n}_2 \otimes \mathbf{n}_1 \otimes \mathbf{n}_2 \otimes \mathbf{n}_1 + \quad (2.9)$$
$$\mathbf{n}_1 \otimes \mathbf{n}_2 \otimes \mathbf{n}_2 \otimes \mathbf{n}_1 + \mathbf{n}_2 \otimes \mathbf{n}_1 \otimes \mathbf{n}_1 \otimes \mathbf{n}_2).$$

Using the explicit representations of the tensors $\mathbf{X}$ and $\mathbb{X}$ and the definition of the double inner product of tensors, we obtain the scalar

$$\mathbf{X} : \mathbf{H} = \mathbf{H} : \mathbb{X} : \mathbf{H} = \mathbf{X} : \tilde{\mathbf{H}} = \tilde{\mathbf{H}} : \mathbb{X} : \tilde{\mathbf{H}} = 2H_{12}^2(s_1 + s_2). \quad (2.10)$$

According to the statements of Proposition 2.1, the necessary and sufficient conditions for the positive definiteness of the tensors $\mathbf{X}$ and $\mathbb{X}$ is the inequality $s_1 + s_2 > 0$, which is confirmed by expression (2.10).

In computational mechanics, second-order tensors are usually represented by column vectors, and fourth-order tensors by matrices. Following this rule, we introduce the following vectors and matrix in Voigt notation

$$[\mathbf{X}] \equiv \begin{bmatrix} X_{11} \\ X_{22} \\ X_{12} \end{bmatrix}, \quad [\mathbf{H}] \equiv \begin{bmatrix} H_{11} \\ H_{22} \\ 2H_{12} \end{bmatrix}, \quad [\mathbf{C}] \equiv \begin{bmatrix} C_{11} & C_{12} & C_{13} \\ C_{21} & C_{22} & C_{23} \\ C_{31} & C_{32} & C_{33} \end{bmatrix},$$

where X_{ij} and H_{ij} ($i, j = 1, 2$) are the components of the symmetric tensors $\mathbf{X}$ and $\mathbf{H}$ in the principal axes of the tensor $\mathbf{S}$, and C_{mn} ($m, n = 1, 2, 3$) are the components of the symmetric matrix $\mathbf{C}$, which are related to the components $\mathbb{X}_{ijkl}$ ($i, j, k, l = 1, 2$) of the tensor $\mathbb{X}$ in the principal axes of the tensor $\mathbf{S}$ by the following expressions:

$$C_{11} = \mathbb{X}_{1111}, \quad C_{12} = \mathbb{X}_{1122}, \quad C_{13} = \mathbb{X}_{1112} = \mathbb{X}_{1121},$$
$$C_{21} = \mathbb{X}_{2211}, \quad C_{22} = \mathbb{X}_{2222}, \quad C_{23} = \mathbb{X}_{2212} = \mathbb{X}_{2221},$$
$$C_{31} = \mathbb{X}_{1211}, \quad C_{32} = \mathbb{X}_{1222}, \quad C_{33} = \mathbb{X}_{1212} = \mathbb{X}_{1221} = \mathbb{X}_{2112} = \mathbb{X}_{2121}.$$

From (2.9) we obtain the values of the non-zero components of the tensor $\mathbb{X}(\mathbf{S})$ in the principal axes of the tensor $\mathbf{S}$

$$\mathbb{X}_{1212} = \mathbb{X}_{1221} = \mathbb{X}_{2112} = \mathbb{X}_{2121} = (s_1 + s_2)/2,$$

that is, the matrix **C** has the following form:

$$[\mathbf{C}] = \begin{bmatrix} 0 & 0 & 0 \\ 0 & 0 & 0 \\ 0 & 0 & (s_1 + s_2)/2 \end{bmatrix}. \quad (2.11)$$

We obtain a vector-matrix counterpart of the tensor expression (2.10) ($[\mathbf{X}] = [\mathbf{C}][\mathbf{H}]$)

$$[\mathbf{H}]^T [\mathbf{X}] = [\mathbf{H}]^T [\mathbf{C}] [\mathbf{H}] = 2H_{12}^2(s_1 + s_2). \quad (2.12)$$

Naturally, the value of the contraction (2.12) in the vector-matrix representation coincide with the value of the same contraction in the tensor representation (2.10).

2.2 Local Body Deformations and Basic Kinematics

Consider the motion of a body Ω in a three-dimensional Euclidean point space, and let $\mathbf{X}$ and $\mathbf{x}$ be the position vectors of some particle $P \in \Omega$ in the *reference* (fixed at time t_0) and *current* (moving at time t) configurations, respectively. Let $\mathbf{F} \equiv \text{Grad}\,\mathbf{x} = \partial \mathbf{x}/\partial \mathbf{X} \in \mathcal{T}^2$ ($J \equiv \det \mathbf{F} > 0$) be the *deformation gradient* (see, e.g., [7]).

We use the *left (symmetric, positive definite, Eulerian) stretch tensor* $\mathbf{V} \equiv \sqrt{\mathbf{F} \cdot \mathbf{F}^T}$ as the main kinematic quantity. Let the eigenindex of the tensor $\mathbf{V}$ be equal to m. The spectral representations (2.1) and (2.2) can be written as

$$\mathbf{V} = \sum_{k=1}^{3} \lambda_k \, \mathbf{n}_k \otimes \mathbf{n}_k = \sum_{i=1}^{m} \lambda_i \mathbf{V}_i,$$

where $0 < \lambda_k < \infty$, $\lambda_k \in \mathbb{R}$ ($k = 1, 2, 3$) are the *principal stretches*.

We define the *Finger strain tensor* (see, e.g., [12]) as

$$\mathbf{e}^{(2)} \equiv \frac{1}{2}(\mathbf{c} - \mathbf{I}) = \frac{1}{2}\sum_{k=1}^{3}(\lambda_k^2 - 1)\,\mathbf{n}_k \otimes \mathbf{n}_k = \frac{1}{2}\sum_{i=1}^{m}(\lambda_i^2 - 1)\mathbf{V}_i, \quad (2.13)$$

where **c** (**B** is the alternative standard notation) is the *left Cauchy–Green deformation tensor*

$$\mathbf{c} \equiv \mathbf{V}^2 = \mathbf{F} \cdot \mathbf{F}^T = \sum_{k=1}^{3} \lambda_k^2 \, \mathbf{n}_k \otimes \mathbf{n}_k = \sum_{i=1}^{m} \lambda_i^2 \mathbf{V}_i. \quad (2.14)$$

2.2 Local Body Deformations and Basic Kinematics

We define the *volume ratio*

$$J \equiv \det \mathbf{F} = \lambda_1 \lambda_2 \lambda_3, \tag{2.15}$$

and the *modified principal stretches*

$$\bar{\lambda}_k \equiv \lambda_k / J^{1/3} \ (k = 1, 2, 3) \ \Rightarrow \ \bar{J} = \bar{\lambda}_1 \bar{\lambda}_2 \bar{\lambda}_3 = 1, \tag{2.16}$$

which correspond to the eigenvalues of the tensor $\bar{\mathbf{V}}$ due to the Richter–Flory multiplicative decomposition [13][6] of the tensor $\mathbf{V}$

$$\mathbf{V} = \bar{\mathbf{V}} \cdot \check{\mathbf{V}}, \quad \bar{\mathbf{V}} \equiv J^{-1/3} \mathbf{V}, \quad \check{\mathbf{V}} \equiv J^{1/3} \mathbf{I}$$

into a unimodular tensor $\bar{\mathbf{V}}$ and a spherical tensor $\check{\mathbf{V}}$, which are responsible for distortional (isochoric) and volumetric (dilatational) deformations, respectively. We define the *modified Finger strain tensor*

$$\bar{\mathbf{e}}^{(2)} \equiv \frac{1}{2}(\bar{\mathbf{V}}^2 - \mathbf{I}) = \frac{1}{2}(\bar{\mathbf{c}} - \mathbf{I}) = \frac{1}{2} \sum_{k=1}^{3} (\bar{\lambda}_k^2 - 1) \mathbf{n}_k \otimes \mathbf{n}_k = \frac{1}{2} \sum_{i=1}^{m} (\bar{\lambda}_i^2 - 1) \mathbf{V}_i,$$

where

$$\bar{\mathbf{c}} \equiv \bar{\mathbf{V}}^2 = J^{-2/3} \mathbf{c}. \tag{2.17}$$

We will further need the *deviator* of the tensor $\bar{\mathbf{e}}^{(2)}$

$$\text{dev}\, \bar{\mathbf{e}}^{(2)} \equiv \bar{\mathbf{e}}^{(2)} - \frac{1}{3} \text{tr}\, \bar{\mathbf{e}}^{(2)} \mathbf{I} \ \Rightarrow \ \text{tr}(\text{dev}\, \bar{\mathbf{e}}^{(2)}) = \text{dev}\, \bar{\mathbf{e}}^{(2)} : \mathbf{I} = 0.$$

It can be shown that

$$\text{dev}\, \bar{\mathbf{e}}^{(2)} = \frac{1}{2} \text{dev}\, \bar{\mathbf{c}}, \quad \text{dev}\, \bar{\mathbf{c}} \equiv \bar{\mathbf{c}} - \frac{1}{3}(\text{tr}\, \bar{\mathbf{c}}) \mathbf{I} \ \Rightarrow \ \text{tr}(\text{dev}\, \bar{\mathbf{c}}) = \text{dev}\, \bar{\mathbf{c}} : \mathbf{I} = 0. \tag{2.18}$$

Hereinafter, we assume that all the tensors $\mathbf{h} \in \mathcal{T}^2$ are sufficiently smooth functions of the monotonically increasing parameter t (time), and we define the *material time derivative* (material rate) of the tensor $\mathbf{h}$: $\dot{\mathbf{h}} \equiv \partial \mathbf{h} / \partial t$. We introduce the *spatial velocity vector* $\mathbf{v}$, the *spatial velocity gradient* $\boldsymbol{\ell}$

$$\mathbf{v} \equiv \dot{\mathbf{x}}, \quad \boldsymbol{\ell} \equiv \text{grad}\, \mathbf{v} = \dot{\mathbf{F}} \cdot \mathbf{F}^{-1},$$

---

[6] It is noted [14] (see also [15]) that this decomposition was first proposed by Richter in the late 1940s (see [16–19]).

the symmetric Eulerian *stretching (strain rate) tensor* $\mathbf{d} \in \mathcal{T}_{\text{sym}}^2$ and the skew-symmetric *vorticity tensor* $\mathbf{w} \in \mathcal{T}_{\text{skew}}^2$ [7]:

$$\boldsymbol{\ell} = \mathbf{d} + \mathbf{w}, \quad \mathbf{d} \equiv \operatorname{sym} \boldsymbol{\ell} = \frac{1}{2}(\boldsymbol{\ell} + \boldsymbol{\ell}^T), \quad \mathbf{w} \equiv \operatorname{skew} \boldsymbol{\ell} = \frac{1}{2}(\boldsymbol{\ell} - \boldsymbol{\ell}^T). \quad (2.19)$$

It can be shown that the tensor $\mathbf{d}$ has the following representation of the form (2.5), (2.6) (see, e.g., [8], Eq. $(115)_2$):

$$\mathbf{d} = \hat{\mathbf{d}} + \tilde{\mathbf{d}}, \quad \hat{\mathbf{d}} \equiv \sum_{i=1}^{m} \frac{\dot{\lambda}_i}{\lambda_i} \mathbf{V}_i, \quad \tilde{\mathbf{d}} \equiv \sum_{i \neq j=1}^{m} \mathbf{V}_i \cdot \mathbf{d} \cdot \mathbf{V}_j. \quad (2.20)$$

Here $\hat{\mathbf{d}}$ and $\tilde{\mathbf{d}}$ are the components of the tensor $\mathbf{d}$ that are coaxial and orthogonal to the tensor $\mathbf{V}$. Note also the validity of the following equality (see, e.g., [21], Eq. (3.72)):

$$\dot{J} = J \operatorname{tr} \mathbf{d}. \quad (2.21)$$

For infinitesimal strains, the following approximate equalities hold (hereinafter, $\boldsymbol{\varepsilon} \equiv \operatorname{sym} \mathrm{D}\mathbf{u}$ is the *infinitesimal strain tensor*, where $\mathbf{u}$ is the displacement vector):

$$\mathbf{e}^{(2)} \approx \bar{\mathbf{e}}^{(2)} \approx \boldsymbol{\varepsilon}, \quad \mathbf{d} \approx \dot{\boldsymbol{\varepsilon}}, \quad \dot{J}/J = \operatorname{tr} \mathbf{d} \approx \operatorname{tr} \dot{\boldsymbol{\varepsilon}}.$$

Remark 2.1 The kinematic tensors $\mathbf{V}$, $\bar{\mathbf{V}}$, $\mathbf{e}^{(2)}$, $\bar{\mathbf{e}}^{(2)}$, $\mathbf{c}$, $\bar{\mathbf{c}}$, and $\mathbf{d}$ are Eulerian objective tensors (cf., [22, 23]).

2.3 Stress Tensors and Their Rates

We define the *(true) Cauchy stress tensor* $\boldsymbol{\sigma}$ and the *(weighted) Kirchhoff stress tensor* $\boldsymbol{\tau}$:

$$\boldsymbol{\tau} \equiv J \boldsymbol{\sigma}. \quad (2.22)$$

We also define the *(engineering, nominal) first Piola–Kirchhoff (1st PK) stress tensor* $\mathbf{P}$:

$$(\mathbf{S}_1 =) \mathbf{P} \equiv \boldsymbol{\tau} \cdot \mathbf{F}^{-T} = J\boldsymbol{\sigma} \cdot \mathbf{F}^{-T} = \boldsymbol{\sigma} \cdot \operatorname{Cof} \mathbf{F}, \quad (2.23)$$

whose components are usually determined in experimental studies. For hyperelastic materials, the 1st PK stress tensor is determined from the elastic energy $W(\mathbf{F})$

$$(\mathbf{S}_1(\mathbf{F}) =) \mathbf{P}(\mathbf{F}) = \mathrm{D}_F W(\mathbf{F}).$$

Remark 2.2 The stress tensors $\boldsymbol{\sigma}$ and $\boldsymbol{\tau}$ are Eulerian objective tensors.

---

[7] Since we assume that the motion law $\mathbf{x}(\mathbf{X}, t) \in C^2$ of t, it follows that $\boldsymbol{\ell}, \mathbf{d}, \mathbf{w} \in C^0$ of t (cf., [20]).

2.4 Elastic Energy and Constitutive Relations for the Linear Isotropic Elastic Model   17

The material rates $\dot{\boldsymbol{\sigma}}$ and $\dot{\boldsymbol{\tau}}$ do not have the Eulerian objectivity property. As the objective rates of the stress tensors we use the Eulerian *Zaremba–Jaumann* (corotational) and *upper Oldroyd* (non-corotational) *stress rates* for the Kirchhoff stress tensor

$$\frac{D^{ZJ}}{Dt}[\boldsymbol{\tau}] \equiv \dot{\boldsymbol{\tau}} + \boldsymbol{\tau}\cdot\mathbf{w} - \mathbf{w}\cdot\boldsymbol{\tau}, \qquad \frac{D^{\overline{Old}}}{Dt}[\boldsymbol{\tau}] \equiv \dot{\boldsymbol{\tau}} - \boldsymbol{\tau}\cdot\boldsymbol{\ell}^T - \boldsymbol{\ell}\cdot\boldsymbol{\tau}, \qquad (2.24)$$

and, likewise, for the Cauchy stress tensor

$$\frac{D^{ZJ}}{Dt}[\boldsymbol{\sigma}] \equiv \dot{\boldsymbol{\sigma}} + \boldsymbol{\sigma}\cdot\mathbf{w} - \mathbf{w}\cdot\boldsymbol{\sigma}, \qquad \frac{D^{\overline{Old}}}{Dt}[\boldsymbol{\sigma}] \equiv \dot{\boldsymbol{\sigma}} - \boldsymbol{\sigma}\cdot\boldsymbol{\ell}^T - \boldsymbol{\ell}\cdot\boldsymbol{\sigma}. \qquad (2.25)$$

These rates are related by the equalities (see, e.g., [24, 25]):

$$\frac{D^{ZJ}}{Dt}[\boldsymbol{\tau}] = J\frac{D^{BH}}{Dt}[\boldsymbol{\sigma}], \qquad \frac{D^{\overline{Old}}}{Dt}[\boldsymbol{\tau}] = J\frac{D^{Tr}}{Dt}[\boldsymbol{\sigma}], \qquad (2.26)$$

where the tensors $\frac{D^{BH}}{Dt}[\boldsymbol{\sigma}]$ and $\frac{D^{Tr}}{Dt}[\boldsymbol{\sigma}]$ are the (Eulerian objective) *Biezeno–Hencky* [26] (or *Hill* [27]) and *Truesdell* (see, e.g., [7]) *stress rates* for the Cauchy stress tensor

$$\frac{D^{BH}}{Dt}[\boldsymbol{\sigma}] \equiv \dot{\boldsymbol{\sigma}} + \boldsymbol{\sigma}\cdot\mathbf{w} - \mathbf{w}\cdot\boldsymbol{\sigma} + \boldsymbol{\sigma}\operatorname{tr}\mathbf{d} \; (= \frac{D^{ZJ}}{Dt}[\boldsymbol{\sigma}] + \boldsymbol{\sigma}\operatorname{tr}\mathbf{d}), \qquad (2.27)$$

$$\frac{D^{Tr}}{Dt}[\boldsymbol{\sigma}] \equiv \dot{\boldsymbol{\sigma}} - \boldsymbol{\sigma}\cdot\boldsymbol{\ell}^T - \boldsymbol{\ell}\cdot\boldsymbol{\sigma} + \boldsymbol{\sigma}\operatorname{tr}\mathbf{d} \; (= \frac{D^{\overline{Old}}}{Dt}[\boldsymbol{\sigma}] + \boldsymbol{\sigma}\operatorname{tr}\mathbf{d}).$$

For infinitesimal strains, the following approximate equalities hold:

$$\boldsymbol{\tau} \approx \mathbf{P} \approx \boldsymbol{\sigma}, \qquad \frac{D^{ZJ}}{Dt}[\boldsymbol{\tau}] \approx \frac{D^{ZJ}}{Dt}[\boldsymbol{\sigma}] \approx \frac{D^{BH}}{Dt}[\boldsymbol{\sigma}] \approx \dot{\boldsymbol{\sigma}}.$$

2.4 Elastic Energy and Constitutive Relations for the Linear Isotropic Elastic Model

Since for infinitesimal strains, the constitutive relations for neo-Hookean models reduce to the constitutive relations for the linear isotropic material model, in this section we provide expressions for the elastic energy and constitutive relations for the latter model as sources inspiring the derivation of corresponding expressions for the former models. In particular, in Sect. 2.4.1, we give expressions for the elastic energy and constitutive relations for the incompressible linear isotropic material model, and in Sects. 2.4.2 and 2.4.3, we present two possible representations of constitutive relations and elastic energy for the compressible linear isotropic material model with coupled (Sect. 2.4.2) and decoupled (Sect. 2.4.3) forms of elastic energy.

2.4.1 Constitutive Relations for Incompressible Materials

The elastic energy for incompressible linear isotropic materials can be written as

$$W_{\text{lin-inc}} \equiv \mu \, \boldsymbol{\varepsilon} : \boldsymbol{\varepsilon} - p \, \text{tr}\, \boldsymbol{\varepsilon} = \mu \|\boldsymbol{\varepsilon}\|^2 - p \, \text{tr}\, \boldsymbol{\varepsilon}, \tag{2.28}$$

where p is the indefinite Lagrange multiplier and $\mu > 0$ is the *shear modulus (first Lamé parameter)*. For these materials, the incompressibility condition holds:

$$\text{tr}\, \boldsymbol{\varepsilon} = 0. \tag{2.29}$$

Note that for infinitesimal strains, $\text{tr}\, \boldsymbol{\varepsilon} \approx J - 1$ holds; i.e., the scalar $\text{tr}\, \boldsymbol{\varepsilon}$ corresponds to the *relative volume ratio (dilatation)*.

The constitutive relations for this material model can be written as

$$\boldsymbol{\sigma} = \frac{\partial W_{\text{lin-inc}}}{\partial \boldsymbol{\varepsilon}} \quad \Leftrightarrow \quad \boldsymbol{\sigma} = 2\mu \, \boldsymbol{\varepsilon} - p \, \mathbf{I}. \tag{2.30}$$

The infinitesimal strain tensor can be represented as

$$\boldsymbol{\varepsilon} = \text{dev}\, \boldsymbol{\varepsilon} + \frac{1}{3} \text{tr}\, \boldsymbol{\varepsilon} \, \mathbf{I}, \quad \text{dev}\, \boldsymbol{\varepsilon} \equiv \boldsymbol{\varepsilon} - \frac{1}{3} \text{tr}\, \boldsymbol{\varepsilon} \, \mathbf{I}. \tag{2.31}$$

Since for incompressible materials, condition (2.29) is satisfied, it follows from (2.31) that $\boldsymbol{\varepsilon} = \text{dev}\, \boldsymbol{\varepsilon}$ for these materials, and from (2.30) we obtain

$$\frac{1}{3} \text{tr}\, \boldsymbol{\sigma} (\equiv \sigma_m) = -p \quad \Rightarrow \quad \sigma_m = -p, \tag{2.32}$$

where σ_m is the *mean stress*. It follows from $(2.32)_2$ that the Lagrange multiplier p for infinitesimal strains of an incompressible isotropic linear elastic material has the mechanical meaning of the *hydrostatic pressure* in the material.

2.4.2 Constitutive Relations for Compressible Materials: Coupled Representation of the Elastic Energy

We represent the elastic energy for linear compressible isotropic materials in the *coupled* form

2.4 Elastic Energy and Constitutive Relations for the Linear Isotropic Elastic Model

$$W_{\text{lin-mixed}} \equiv \mu \, \boldsymbol{\varepsilon} : \boldsymbol{\varepsilon} + \frac{\lambda}{2} \text{tr}^2 \boldsymbol{\varepsilon} = \mu \, \|\boldsymbol{\varepsilon}\|^2 + \frac{\lambda}{2} \text{tr}^2 \boldsymbol{\varepsilon}, \tag{2.33}$$

where λ is the *second Lamé parameter*. We call this form coupled since the summands on the r.h.s. of (2.33) are not decoupled into volumetric and isochoric strains; i.e., the second summand on the r.h.s. of (2.33) depends only on volumetric strain, but the first summand depends on both volumetric and isochoric strains.

The constitutive relations for this material model can be written as

$$\boldsymbol{\sigma} = \frac{\partial W_{\text{lin-mixed}}}{\partial \boldsymbol{\varepsilon}} \quad \Leftrightarrow \quad \boldsymbol{\sigma} = 2\mu \, \boldsymbol{\varepsilon} + \lambda \, \text{tr}\, \boldsymbol{\varepsilon} \, \mathbf{I}. \tag{2.34}$$

Note the formal similarity between the right-hand sides of (2.30)$_2$ and (2.34)$_2$ under the following identifications: $-p\mathbf{I} \leftrightarrow \lambda \, \text{tr}\, \boldsymbol{\varepsilon} \, \mathbf{I}$. However, the mean stress now depends not only on λ, but also on μ due to the equality

$$\sigma_m = (\lambda + \frac{2}{3}\mu) \, \text{tr}\, \boldsymbol{\varepsilon} = K \, \text{tr}\, \boldsymbol{\varepsilon}, \tag{2.35}$$

where $K \equiv \lambda + \frac{2}{3}\mu$ is the *bulk modulus*.

2.4.3 Constitutive Relations for Compressible Materials: Decoupled Representation of the Elastic Energy

The *decoupled* representations of the elastic energy and constitutive relations are obtained from the expressions in Sect. 2.4.2 using the additive separation of the Cauchy strain tensor into the deviatoric and spherical components in the form (2.31)$_1$. In this representation, the elastic energy (2.33) is rewritten as

$$W_{\text{lin-vol-iso}}(= W_{\text{lin-mixed}}) \equiv \mu \, \text{dev}\, \boldsymbol{\varepsilon} : \text{dev}\, \boldsymbol{\varepsilon} + \frac{K}{2}(\text{tr}\, \boldsymbol{\varepsilon})^2 = \mu \, \|\text{dev}\, \boldsymbol{\varepsilon}\|^2 + \frac{K}{2}\text{tr}^2 \boldsymbol{\varepsilon}. \tag{2.36}$$

The second summand on the r.h.s. of (2.36) corresponds to volumetric strain, and the first summand corresponds to isochoric strain.

The constitutive relations (2.34) can be rewritten as

$$\boldsymbol{\sigma} = \frac{\partial W_{\text{lin-vol-iso}}}{\partial \boldsymbol{\varepsilon}} \quad \Leftrightarrow \quad \boldsymbol{\sigma} = 2\mu \, \text{dev}\, \boldsymbol{\varepsilon} + K \, \text{tr}\, \boldsymbol{\varepsilon} \, \mathbf{I}, \tag{2.37}$$

whence we obtain the expression for the mean stress

$$\sigma_m (\equiv \frac{1}{3}\text{tr}\, \boldsymbol{\sigma}) = K \, \text{tr}\, \boldsymbol{\varepsilon},$$

which naturally coincides with (2.35).

References

1. Itskov, M.: Tensor Algebra and Tensor Analysis for Engineers (with Applications to Continuum Mechanics), 5th edn. Springer, Cham (2019)
2. Federico, S.: Covariant formulation of the tensor algebra of non-linear elasticity. Int. J. Non-Linear Mech. **47**(2), 273–284 (2012). https://doi.org/10.1016/j.ijnonlinmec.2011.06.007
3. Federico, S., Holthausen, S., Husemann, N.J., Neff, P.: Major symmetry of the induced tangent stiffness tensor for the Zaremba-Jaumann rate and Kirchhoff stress in hyperelasticity: Two different approaches. Math. Mech. Solids (2025). https://doi.org/10.1177/10812865241306703. (in press)
4. Curnier, A.: Computational Methods in Solid Mechanics. Kluwer, Dordrecht (1994)
5. Holzapfel, G.A.: Nonlinear Solid Mechanics: A Continuum Approach for Engineering. Wiley, Chichester (2000)
6. Peyraut, F., Feng, Z.Q., He, Q.C., Labed, N.: Robust numerical analysis of homogeneous and non-homogeneous deformations. Appl. Numer. Math. **59**(7), 1499–1514 (2009). https://doi.org/10.1016/j.apnum.2008.10.002
7. Bertram, A.: Elasticity and Plasticity of Large Deformations, 4th edn. Springer, Cham (2021)
8. Korobeynikov, S.N.: Families of continuous spin tensors and applications in continuum mechanics. Acta Mech. **216**(1–4), 301–332 (2011). https://doi.org/10.1007/s00707-010-0369-7
9. Korobeynikov, S.N.: Discussion of "The general basis-free spin and its concise proof" by Meng and Chen. Acta Mech. **234**(2), 825–829 (2023). https://doi.org/10.1007/s00707-022-03402-4
10. Luehr, C.P., Rubin, M.B.: The significance of projection operators in the spectral representation of symmetric second order tensors. Comput. Methods Appl. Mech. Eng. **84**(3), 243–246 (1990). https://doi.org/10.1016/0045-7825(90)90078-Z
11. Korobeynikov, S.N.: Basis-free expressions for families of objective strain tensors, their rates, and conjugate stress tensors. Acta Mech. **229**(3), 1061–1098 (2018). https://doi.org/10.1007/s00707-017-1972-7
12. Curnier, A., Rakotomanana, L.: Generalized strain and stress measures: critical survey and new results. Eng. Trans. **39**(3–4), 461–538 (1991)
13. Flory, P.J.: Thermodynamic relations for high elastic materials. Trans. Faraday Soc. **57**, 829–838 (1961). https://doi.org/10.1039/TF9615700829
14. Graban, K., Schweickert, E., Martin, R.J., Neff, P.: A commented translation of Hans Richter's early work "The isotropic law of elasticity." Math. Mech. Solids **24**(8), 2649–2660 (2019). https://doi.org/10.1177/1081286519847495
15. Neff, P., Holthausen, S., d'Agostino, M.V., Bernardini, D., Sky, A., Ghiba, I.D., Martin, R.J.: Hypo-elasticity, Cauchy-elasticity, corotational stability and monotonicity in the logarithmic strain. J. Mech. Phys. Solids **202**, 106074 (2025). https://doi.org/10.1016/j.jmps.2025.106074
16. Richter, H.: Das isotrope Elastizitätsgesetz. Z. Angew. Math. Mech. **28**(7–8), 205–209 (1948). https://doi.org/10.1002/zamm.19480280703
17. Richter, H.: Hauptaufsätze: Verzerrungstensor, Verzerrungsdeviator und Spannungstensor bei endlichen Formänderungen. Z. Angew. Math. Mech. **29**(3), 65–75 (1949). https://doi.org/10.1002/zamm.19490290301
18. Richter, H.: Zum Logarithmus einer Matrix. Arch. Math. **2**(5), 360–363 (1949). https://doi.org/10.1007/BF02036865
19. Richter, H.: Zur Elastizitätstheorie endlicher Verformungen. Mathematische Nachrichten **8**(1), 65–73 (1952). https://doi.org/10.1002/mana.19520080109
20. Scheidler, M.: Time rates of generalized strain tensors Part I: Component formulas. Mech. Mater. **11**(3), 199–210 (1991). https://doi.org/10.1016/0167-6636(91)90002-H
21. de Souza Neto, E.A., Peric, D., Owen, D.J.R.: Computational Methods for Plasticity: Theory and Applications. Wiley, Chichester (2008)
22. Korobeynikov, S.N.: Objective tensor rates and applications in formulation of hyperelastic relations. J. Elast. **93**(2), 105–140 (2008). https://doi.org/10.1007/s10659-008-9166-0

References

23. Ogden, R.W.: Non-linear Elastic Deformations. Ellis Horwood, Chichester (1984)
24. Korobeynikov, S., Larichkin, A.: Objective Algorithms for Integrating Hypoelastic Constitutive Relations Based on Corotational Stress Rates. Springer, Cham (2023)
25. Korobeynikov, S.N., Larichkin, A.Y.: Simulating body deformations with initial stresses using Hooke-like isotropic hypoelasticity models based on corotational stress rates. Z. Angew. Math. Mech. **104**(2), e202300568 (2024). https://doi.org/10.1002/zamm.202300568
26. Biezeno, C., Hencky, H.: On the general theory of elastic stability. Koninklijke Akademie van Wettenschappen te Amsterdam **31**, 569–592 (1928)
27. Hill, R.: A general theory of uniqueness and stability in elastic-plastic solids. J. Mech. Phys. Solids **6**(3), 236–249 (1958). https://doi.org/10.1016/0022-5096(58)90029-2

Chapter 3
Constitutive Relations for Neo-Hookean Isotropic Hyperelastic Material Models

Abstract In this chapter we present constitutive relations for three types of *neo-Hookean isotropic hyperelastic material models*: incompressible (classical) model (Sect. 3.1), compressible mixed models (Sect. 3.2), and compressible vol-iso models (Sect. 3.3). The formulations of these three types of models are discussed in Sect. 3.4. Although all three types of models are presented in one form or another in the literature (see Chap. 1), the objectives of the present chapter is to unify them in the Eulerian formulation and relate them to the linear isotropic elastic models formulated in Sect. 2.4.

3.1 Constitutive Relations for the Incompressible Neo-Hookean Isotropic Hyperelastic Material Model

We represent the incompressible *neo-Hookean material model* as the one-power generalized Ogden material model (cf., [1]) for $n = 2$ with the modified expression for the elastic energy (cf., [2])

$$W_{\text{neo-Hooke}}(\lambda_1, \lambda_2, \lambda_3) \equiv \mu\,(\text{tr}\,\mathbf{e}^{(2)} - \ln J) - p \ln J = \frac{\mu}{2}(\|\mathbf{F}\|^2 - 3 - 2\ln J) - p \ln J, \tag{3.1}$$

where the parameter p is the indefinite Lagrange multiplier. According to (2.13), the quantity $\text{tr}\,\mathbf{e}^{(2)}$ is defined as

$$\text{tr}\,\mathbf{e}^{(2)} = \frac{1}{2}(\text{tr}\,\mathbf{c} - 3) = \frac{1}{2}(\lambda_1^2 + \lambda_2^2 + \lambda_3^2 - 3) = \frac{1}{2}(\|\mathbf{F}\|^2 - 3). \tag{3.2}$$

Since $\ln J = \ln \lambda_1 + \ln \lambda_2 + \ln \lambda_3$, the potential function (3.1) satisfies the Valanis–Landel hypothesis [3, 4], i.e.,

$$W_{\text{neo-Hooke}}(\lambda_1, \lambda_2, \lambda_3) = \sum_{k=1}^{3} \check{W}_{\text{neo-Hooke}}(\lambda_k),$$

$$\check{W}_{\text{neo-Hooke}}(\lambda) \equiv \mu \left[\frac{1}{2}(\lambda^2 - 1) - \ln \lambda\right] - p \ln \lambda.$$

It has been shown [2] that for infinitesimal strains, the elastic energy (3.1) reduces to the elastic energy (2.28).

Since for isotropic hyperelastic materials, the principal directions of the tensors σ and $\mathbf{V}$ (and $\mathbf{c}$) coincide (see, e.g., [5]), the Cauchy stress tensor σ can be represented as

$$\boldsymbol{\sigma} = \sum_{k=1}^{3} \sigma_k \, \mathbf{n}_k \otimes \mathbf{n}_k = \sum_{i=1}^{m} \sigma_i \mathbf{V}_i,$$

where σ_k ($k = 1, 2, 3$) or σ_i ($i = 1, \ldots, m$) are the *principal Cauchy stresses* defined for incompressible materials as follows (see, e.g., [1, 6, 7]):

$$\sigma_k = \lambda_k \frac{\partial W_{\text{neo-Hooke}}}{\partial \lambda_k} \ (k = 1, 2, 3) \quad \Leftrightarrow \quad \sigma_i = \lambda_i \frac{\partial W_{\text{neo-Hooke}}}{\partial \lambda_i} \ (i = 1, \ldots, m). \tag{3.3}$$

Basis-free expressions for the stress tensor σ can be obtained from (3.1), (3.2), and (3.3) [2]

$$\boxed{\boldsymbol{\sigma} = 2\mu \, \mathbf{e}^{(2)} - p\,\mathbf{I} = \mu\,(\mathbf{c} - \mathbf{I}) - p\,\mathbf{I}}. \tag{3.4}$$

It has been shown [2] that for infinitesimal strains, the constitutive relations (3.4) reduce to the constitutive relations $(2.30)_2$ for linear elastic materials.

To determine the mean stress σ_m, we rewrite the constitutive relations (3.4) as

$$\boldsymbol{\sigma} = 2\mu \, \text{dev}\,\mathbf{e}^{(2)} + \frac{2\mu}{3} \text{tr}\,\mathbf{e}^{(2)}\,\mathbf{I} - p\,\mathbf{I}, \tag{3.5}$$

where $\text{dev}\,\mathbf{e}^{(2)}$ is the Finger strain tensor deviator defined as

$$\text{dev}\,\mathbf{e}^{(2)} \equiv \mathbf{e}^{(2)} - \frac{1}{3} \text{tr}\,\mathbf{e}^{(2)}\,\mathbf{I}.$$

The mean stress σ_m ($\equiv \text{tr}\,\boldsymbol{\sigma}/3$) can be obtained from (3.5):

$$\sigma_m = -p + \frac{2}{3}\mu\,\text{tr}\,\mathbf{e}^{(2)}. \tag{3.6}$$

3.2 Constitutive Relations for the Compressible Mixed Isotropic Hyperelastic Neo-Hookean Material Models

Note that unlike the linear elastic incompressible isotropic material model, the Lagrange multiplier p for the neo-Hookean material model can no longer be interpreted as the hydrostatic pressure $(-\sigma_m)$. However, for infinitesimal strains,

$$\operatorname{tr} \mathbf{e}^{(2)} \approx \operatorname{tr} \boldsymbol{\varepsilon} \approx 0,$$

and in this case, equality (3.6) becomes equality (2.32); i.e., $\sigma_m = -p$.

3.2 Constitutive Relations for the Compressible Mixed Isotropic Hyperelastic Neo-Hookean Material Models

We use the following form of the elastic energy for any *compressible mixed (vol-iso coupled) neo-Hookean* material model (a *mixed* material model) (cf., [8–11]):

$$W_{\text{mixed}}(\lambda_1, \lambda_2, \lambda_3) \equiv \mu(\operatorname{tr} \mathbf{e}^{(2)} - \ln J) + \lambda\, h(J) = \frac{\mu}{2}(\|\mathbf{F}\|^2 - 3 - 2\ln J) + \lambda\, h(J). \tag{3.7}$$

Here $h(J)$ is a scalar *volumetric function* of J, whose properties and forms will be considered below[1] and the quantity $\operatorname{tr} \mathbf{e}^{(2)}$ is defined in (3.2).

For isotropic hyperelastic materials, the Kirchhoff stress tensor is represented as

$$\boldsymbol{\tau} = \sum_{k=1}^{3} \tau_k\, \mathbf{n}_k \otimes \mathbf{n}_k = \sum_{i=1}^{m} \tau_i\, \mathbf{V}_i, \tag{3.8}$$

i.e., the principal axes of the tensors $\boldsymbol{\tau}$ and $\mathbf{V}$ coincide. The constitutive relations for this material model can be written as relations between the principal Kirchhoff stresses and the principal stretches (see, e.g., [6, 7])

$$\tau_k = \lambda_k \frac{\partial W_{\text{mixed}}}{\partial \lambda_k} \ (k=1,2,3) \quad \Leftrightarrow \quad \tau_i = \lambda_i \frac{\partial W_{\text{mixed}}}{\partial \lambda_i} \ (i=1,\ldots,m). \tag{3.9}$$

Based on (3.8) and (3.9), the Kirchhoff stress tensor can be represented in terms of kinematic quantities by the following basis-free expression (for details, see [12]):

$$\boldsymbol{\tau} = 2\mu\, \mathbf{e}^{(2)} + \lambda\, J h'(J) \mathbf{I} = \mu\, (\mathbf{c} - \mathbf{I}) + \lambda\, J h'(J) \mathbf{I}, \tag{3.10}$$

where

$$h'(J) \equiv \frac{d\, h(J)}{d\, J}. \tag{3.11}$$

---

[1] We now require that for infinitesimal strains, $h(J) \approx \frac{1}{2}(J-1)^2 \approx \frac{1}{2}\operatorname{tr}^2 \boldsymbol{\varepsilon}$.

Based on (2.22) and (3.10), the Cauchy stress tensor can be represented by the basis-free expression

$$\sigma = \frac{2\mu}{J}\mathbf{e}^{(2)} + \lambda h'(J)\mathbf{I} = \frac{\mu}{J}(\mathbf{c} - \mathbf{I}) + \lambda h'(J)\mathbf{I}. \qquad (3.12)$$

3.3 Constitutive Relations for the Compressible vol-iso Isotropic Hyperelastic Neo-Hookean Material Models

Following [13], we write the elastic energy for any *compressible vol-iso neo-Hookean* material model (the *vol-iso* material model) as

$$W_{\text{vol-iso}}(\lambda_1, \lambda_2, \lambda_3) \equiv \mu\,(\operatorname{tr}\bar{\mathbf{e}}^{(2)} - \ln \bar{J}) + Kh(J) = \frac{\mu}{2}(\|\frac{\mathbf{F}}{J^{1/3}}\|^2 - 3 - 2\ln \bar{J}) + Kh(J), \qquad (3.13)$$

The principal components of the Kirchhoff stress tensor τ are determined using expressions (3.9) with the replacement of elastic energy W_{mixed} by $W_{\text{vol-iso}}$. As a result, we obtain the following basis-free expression for the Kirchhoff stress tensor (for details, see [13]):

$$\tau = 2\mu\operatorname{dev}\bar{\mathbf{e}}^{(2)} + KJh'(J)\mathbf{I} = \mu\operatorname{dev}\bar{\mathbf{c}} + KJh'(J)\mathbf{I}, \qquad (3.14)$$

here we also used equality (2.18)$_1$. Equations (2.22) and (3.14) lead to the following expression for the Cauchy stress tensor:

$$\sigma = \frac{2\mu}{J}\operatorname{dev}\bar{\mathbf{e}}^{(2)} + Kh'(J)\mathbf{I} = \frac{\mu}{J}\operatorname{dev}\bar{\mathbf{c}} + Kh'(J)\mathbf{I}. \qquad (3.15)$$

Remark 3.1 We can show the validity of the equalities

$$\frac{\partial \bar{J}}{\partial \lambda_1} = \frac{\partial \bar{J}}{\partial \lambda_2} = \frac{\partial \bar{J}}{\partial \lambda_3} = 0,$$

whence the first summand on the r.h.s. of (3.13) can be rewritten as $\mu\operatorname{tr}\bar{\mathbf{e}}^{(2)}$, but expression (3.14) for the Kirchhoff stress tensor τ does not change in this case. We retained the term $\ln \bar{J}$ in expression (3.13) to emphasize that this expression for compressible materials is similar to expression (3.1) for incompressible materials.

3.4 Discussion of Expressions for Compressible Neo-Hookean Material Models

For infinitesimal strains, the following equality holds:

$$\ln J \approx J - 1 \approx \operatorname{tr} \boldsymbol{\varepsilon}.$$

It has been shown [2] that for these strains, expression (3.1) for the elastic energy reduces to expression (2.28) and expression (3.4) for constitutive relations reduces to expression (2.30)$_2$. In addition, for small strains, the following approximations are valid:

$$h(J) \approx \frac{1}{2}(\ln J)^2 \approx \frac{1}{2}(\operatorname{tr} \boldsymbol{\varepsilon})^2 \quad \Rightarrow \quad h'(J) \approx J h'(J) \approx \ln J \approx \operatorname{tr} \boldsymbol{\varepsilon}.$$

Then for infinitesimal strains, expression (3.7) for the elastic energy for any mixed material model becomes expression (2.33), and expression (3.13) for the elastic energy for any vol-iso material model becomes expression (2.36). In addition, for these strains, expressions (3.10) and (3.12) for constitutive relations become expression (2.34)$_2$, and expressions (3.14) and (3.15) for constitutive relations become expression (2.37)$_2$.

Since, within the framework of linear elasticity theory, both the elastic energy expressions (2.33) and (2.36) and the constitutive relations (2.34) and (2.37) are equivalent, the mixed and vol-iso models for infinitesimal strains reduce to the same linear compressible isotropic elastic material model. However, in the general case of strains of arbitrary magnitude, the mixed and vol-iso models are different models of compressible isotropic hyperelastic materials. In particular, the elastic energy (3.7) for any mixed material model is an additive decomposition into an uncoupled component (first summand on the r.h.s. of (3.7)) and a coupled component (second summand on the r.h.s. of (3.7)) with respect to λ_k ($k = 1, 2, 3$) (cf., [14]). That is, the elastic energy for any mixed model is consistent with the Valanis–Landel hypothesis [4] extended to compressible materials in [3].[2] At the same time, the elastic energy for any vol-iso model is inconsistent with this hypothesis; however, the elastic energy (3.13) for this model corresponds to the additive decomposition into a volumetric component (second summand on the right-hand side of (3.13)) and an isochoric component (first summand on the right-hand side of (3.13)). That is, any vol-iso model allows the uncoupled representation of the elastic energy as a sum of volumetric and isochoric strains, but the mixed model does not allow this representation. With regard to the above decompositions in the elastic energy expressions, we cannot give preference to either one of these two models, since both types of decomposition have not been confirmed experimentally. In particular, experimental studies by Vangerko and Treloar [15] have shown that the Valanis–

---

[2] Valanis [3] requires such decomposition of the elastic energy into uncoupled and coupled components corresponding to the mixed (isochoric and volumetric) and pure volumetric energies.

Landel hypothesis is not valid for sufficiently large values of the principal stretches ($\gtrsim 3$). At the same time, it has been shown [16] that the *additive decomposition of the elastic energy into any volumetric and isochoric parts is inconsistent* with experimental data on the dependence of the dilatation $J - 1$ on longitudinal extension under uniaxial loading.

Next, along with the Lamé parameters λ and μ, we will also use more physically reasonable parameters—*Young's modulus E* and *Poisson's ratio* ν, which are related to the Lamé parameters as follows (see, e.g., Table 5 in [17]):

$$\mu = \frac{E}{2(1+\nu)}, \quad \lambda = \frac{E\nu}{(1+\nu)(1-2\nu)} \quad \Leftrightarrow \quad E = \frac{\mu(3\lambda+2\mu)}{\lambda+\mu}, \quad \nu = \frac{\lambda}{2(\lambda+\mu)}. \tag{3.16}$$

We now perform a preliminary analysis to determine admissible values of the parameters λ and μ for the mixed and vol-iso models. Since for infinitesimal strains, both types of models reduce to the linear elastic compressible material model, we use the well-known constraints on the parameters K and μ (see, e.g., [18])

$$\mu > 0, \quad K = \lambda + 2\mu/3 > 0, \tag{3.17}$$

under which the potential energy $W_{\text{lin-vol-iso}}$ in (2.36) is positive definite, i.e., $\varepsilon \to W_{\text{lin}}(\varepsilon)$ is strictly convex. In view of expressions (3.16), the constraints (3.17) can be represented in terms of μ and ν:

$$\mu > 0, \quad -1 < \nu \leq 0.5. \tag{3.18}$$

For isotropic hyperelastic models with elastic energies of the form (3.13), the convexity condition will be imposed on the function $Kh(J)$ [19]. Since we assume that $K > 0$ (see (3.17)), we further consider convex functions $h(J)$. Next, we use identical functions $h(J)$ for both material models (mixed and vol-iso) and relax the constraint on the function $\lambda h(J)$ for the mixed model: we replace the convexity condition by the requirement of non-concavity; i.e., for any mixed model, the parameters λ and μ (or ν and μ) are subjected to the constraints

$$\mu > 0, \lambda \geq 0 \quad \Leftrightarrow \quad \mu > 0, \quad 0 \leq \nu \leq 0.5, \tag{3.19}$$

while for any vol-iso model, the constraints on the parameters ν and μ are still given by (3.18). Note that by our convention, in both constraints (3.18) and (3.19), the value $\nu = 0.5$ means the use of the classical neo-Hookean model for incompressible materials instead of the mixed and vol-iso models.

Particular attention should be given to the performance analysis of the mixed and vol-iso models for slightly compressible materials characterized by Poisson's ratio $\nu \approx 0.5^-$ (e.g., $\nu = 0.4999$). First, all elastomers in practice are slightly compressible materials, and, second, even if a researcher wants to perform computer simulations of deformations of incompressible materials using some commercial

3.4 Discussion of Expressions for Compressible Neo-Hookean Material Models

FE systems, he/she will be faced with the fact that these systems use slightly compressible approximations, rather than purely incompressible materials, to impose the incompressibility condition by means of the penalty function method. Ideally, the mixed and vol-iso compressible material models in the case of slight compressibility should lead to similar solutions for stresses and kinematic quantities that approximate the corresponding solutions for incompressible materials. We now find constraints on the model parameters that should lead to this result.

We assume that the principal stretches λ_k ($k = 1, 2, 3$) for slightly compressible materials are similar to those for incompressible materials. Since in the last case, $J = 1$, it follows that for slightly compressible materials, the approximate equality $J \approx 1$ should hold. In this case, (2.16) lead to the approximate equalities $\bar{\lambda}_k \approx \lambda_k$ ($k = 1, 2, 3$). Hence for slightly compressible materials, the first terms on the right-hand sides of (3.7) and (3.13) should approximate the first term on the r.h.s. of (3.1).

Since for slightly compressible materials, the *dilatation* $J - 1$ is small, functions $h(J)$ satisfy the approximations (see Chap. 4):

$$h(J) \approx \frac{1}{2}(\ln J)^2 \approx \frac{1}{2}(J - 1)^2.$$

We assume that the potential energies of volumetric and distortional strains have comparable values. Then in view of (3.7) and (3.13), the quantities λ and K should far exceed the shear modulus. It follows from (3.16) that this condition should be satisfied for values of Poisson's ratio close to 0.5. Typically, in computer simulations of deformations of incompressible materials, it is assumed that $\nu = 0.4999$. In this case,

$$\frac{\lambda}{\mu} = \frac{2\nu}{1 - 2\nu} \approx 5000;$$

i.e., $\lambda \approx K \approx 5000\mu$.

We assume that for solutions of boundary-value problems of deformations for slightly compressible hyperelastic materials,

$$\lim_{\nu \to 0.5^-} \frac{\lambda(\nu)}{2} \ln J(\nu) = -p, \tag{3.20}$$

which is in fact equivalent to the application of the penalty function method to the solution of deformation problems for incompressible hyperelastic materials. Comparing the first terms on the right-hand sides of (3.7) and (3.13) with the first term on the right-hand side of (3.1), we conclude that simulations of deformations using both mixed and vol-iso models in the case of slight compressibility approximate similar simulations using the neo-Hookean model provided that equality (3.20) holds.

Remark 3.2 Note that for slightly compressible materials, both types of models (mixed and vol-iso) can only be used to determine the Cauchy stress tensor σ and the principal stretches λ_k ($k = 1, 2, 3$). However, the use of the obtained values of the principal stretches generally does not lead to a correct determination of the volume ratio from Eq. (2.15), as shown by the experimental studies of Penn [16]. The reason for the disagreement between the values of the quantity J obtained using standard compressible material models and experimental data is the difference in the scales of mechanical quantities; i.e., although the values of λ_k ($k = 1, 2, 3$) can differ greatly from 1, their product in (2.15) is close to 1. Various modifications of elastic energies for vol-iso models have been proposed to improve their performance compared to experimental data (see e.g., [7, 20–26]). However, the performance analysis of these models is beyond the scope of this study.

Remark 3.3 Our conclusion that in simulations of deformations, solutions for stresses and principal stretches for slightly compressible materials are close to those for incompressible neo-Hookean materials refer only to finite values of the principal stretches ($0.1 \lesssim \lambda_k \lesssim 10$) ($k = 1, 2, 3$), which is quite sufficient for applications. However, for extreme values of λ_k ($\lambda_k \to 0$ and $\lambda_k \to \infty$), such closeness of solutions is not always the case. The behavior of solutions for extreme values of λ_k ($k = 1, 2, 3$) is heavily affected by the choice of the function $h(J)$ (see Chap. 6).

References

1. Ogden, R.W.: Large deformation isotropic elasticity—on the correlation of theory and experiment for incompressible rubberlike solids. Proceedings of the Royal Society of London. A. Math. Phys. Sci. **326**(1567), 565–584 (1972). https://doi.org/10.1098/rspa.1972.0026
2. Korobeynikov, S.N.: Generalized Ogden's incompressible isotropic hyperelastic material model supporting the form of Hooke's law. Arch. Appl. Mech. **95**(5), 105 (2025). https://doi.org/10.1007/s00419-025-02815-6
3. Valanis, K.C.: The Valanis-Landel strain energy function elasticity of incompressible and compressible rubber-like materials. Int. J. Solids Struct. **238**, 111271 (2022). https://doi.org/10.1016/j.ijsolstr.2021.111271
4. Valanis, K.C., Landel, R.F.: The strain-energy function of a hyperelastic material in terms of the extension ratios. J. Appl. Phys. **38**(7), 2997–3002 (1967). https://doi.org/10.1063/1.1710039
5. Bertram, A.: Elasticity and Plasticity of Large Deformations, 4th edn. Springer, Cham (2021)
6. Korobeynikov, S.N.: Objective symmetrically physical strain tensors, conjugate stress tensors, and Hill's linear isotropic hyperelastic material models. J. Elasticity **136**(2), 159–187 (2019). https://doi.org/10.1007/s10659-018-9699-9
7. Ogden, R.W.: Non-linear Elastic Deformations. Ellis Horwood, Chichester (1984)
8. Korobeynikov, S.N.: Families of Hooke-like isotropic hyperelastic material models and their rate formulations. Arch. Appl. Mech. **93**(10), 3863–3893 (2023). https://doi.org/10.1007/s00419-023-02466-5
9. Ogden, R.W.: Large deformation isotropic elasticity: on the correlation of theory and experiment for compressible rubberlike solids. Proceedings of the Royal Society of London. A. Mathematical and Physical Sciences **328**(1575), 567–583 (1972). https://doi.org/10.1098/rspa.1972.0096
10. Simo, J.C., Pister, K.S.: Remarks on rate constitutive equations for finite deformation problems: computational implications. Comput. Methods Appl. Mech. Eng. **46**(2), 201–215 (1984). https://doi.org/10.1016/0045-7825(84)90062-8
11. Wriggers, P.: Nonlinear Finite Element Methods. Springer, Berlin, Heidelberg (2008)

References

12. Korobeynikov, S.N.: Family of continuous strain-consistent convective tensor rates and its application in Hooke-like isotropic hypoelasticity. J. Elasticity **143**(1), 147–185 (2021). https://doi.org/10.1007/s10659-020-09808-2
13. de Souza Neto, E.A., Peric, D., Owen, D.J.R.: Computational Methods for Plasticity: Theory and Applications. Wiley, Chichester (2008)
14. Kellermann, D.C., Attard, M.M.: An invariant-free formulation of neo-Hookean hyperelasticity. Zeitschrift für Angewandte Mathematik und Mechanik **96**(2), 233–252 (2016). https://doi.org/10.1002/zamm.201400210
15. Vangerko, H., Treloar, L.R.G.: The inflation and extension of rubber tube for biaxial strain studies. J. Phys. D: Appl. Phys. **11**(14), 1969–1978 (1978). https://doi.org/10.1088/0022-3727/11/14/009
16. Penn, R.W.: Volume changes accompanying the extension of rubber. Trans. Soc. Rheol. **14**, 509–517 (1970). https://doi.org/10.1122/1.549176
17. Rubin, M.B.: Continuum Mechanics with Eulerian Formulations of Constitutive Equations. Springer, Cham (2021)
18. Batra, R.C.: Elements of Continuum Mechanics. AIAA, Reston (2006)
19. Hartmann, S., Neff, P.: Polyconvexity of generalized polynomial-type hyperelastic strain energy functions for near-incompressibility. Int. J. Solids Struct. **40**(11), 2767–2791 (2003). https://doi.org/10.1016/S0020-7683(03)00086-6
20. Attard, M.M.: Finite strain-isotropic hyperelasticity. Int. J. Solids Struct. **40**(17), 4353–4378 (2003). https://doi.org/10.1016/S0020-7683(03)00217-8
21. Attard, M.M., Hunt, G.W.: Hyperelastic constitutive modeling under finite strain. Int. J. Solids Struct. **41**(18), 5327–5350 (2004). https://doi.org/10.1016/j.ijsolstr.2004.03.016
22. Fong, J.T., Penn, R.W.: Construction of a strain-energy function for an isotropic elastic material. Trans. Soc. Rheol. **19**(1), 99–113 (1975). https://doi.org/10.1122/1.549389
23. Huang, Z.P.: A novel constitutive formulation for rubberlike materials in thermoelasticity. J. Appl. Mech. **81**(4), 041013 (2014). https://doi.org/10.1115/1.4025272
24. Huang, Z.P.: Erratum: "A novel constitutive formulation for rubberlike materials in thermoelasticity" [asme j. appl. mech., 2014, 81(4), p. 041013]. Journal of Applied Mechanics **83**(4), 047001 (2016). https://doi.org/10.1115/1.4032660
25. Rogovoy, A.: Effect of elastomer slight compressibility. European J. Mech.–A/Solids **20**(5), 757–775 (2001). https://doi.org/10.1016/S0997-7538(01)01170-6
26. Yao, Y., Chen, S., Huang, Z.: A generalized Ogden model for the compressibility of rubberlike solids. Philosophical Transactions of the Royal Society A: Mathematical, Physical and Engineering Sciences **380**(2234), 20210320 (2022). https://doi.org/10.1098/rsta.2021.0320

Chapter 4
Some Volumetric Functions and Their Properties

Abstract The objective of this chapter is to summarize properties of volumetric functions available in the literature (Sect. 4.1) and to analyze the performance of eight such functions widely presented in the literature (Sect. 4.2).

4.1 Properties of Volumetric Functions

Following [1, 2], we impose the following constraints on the function $h(J) \in C^2$ ($0 < J < \infty$); i.e., we require it to have the following properties:

1. Properties of the function $h(J)$ at the point $J = 1$ [1, 2]

$$h(1) = 0, \quad h' \equiv \frac{d\,h(J)}{d\,J})|_{J=1} = 0, \quad h'' \equiv \frac{d^2\,h(J)}{d\,J^2})|_{J=1} = 1.$$

2. Properties of the function h' [2]

$$h'|_{J<1} < 0, \quad h'|_{J>1} > 0.$$

3. Rank-one convexity of the function $\mathbf{F} \to h(\det \mathbf{F})$, equivalent to the convexity of $J \to h(J)$ (desired property of the function $h(J)$) [1]

$$h''(J) > 0 \quad \forall\, 0 < J < \infty.$$

4. Hill's stability condition applied to a purely volumetric function for compressible materials [2]

$$\chi(J) \equiv h'(J) + J h''(J) > 0 \quad \forall\, 0 < J < \infty. \tag{4.1}$$

5. Desired properties of the function $h(J)$ in extreme states [1]

$$h(J) \to \infty \text{ for } J \to 0 \text{ and } J \to \infty.$$

Remark 4.1 When replacing the variable J with $\log J$ for the function $h(J)$, we obtain a modified function [3]

$$\tilde{h}(\log J) := h(J).$$

It is easy to show the following equality

$$\chi(J) = \frac{\partial^2 \tilde{h}(\log J)}{\partial (\log J)^2} \frac{1}{J},$$

from which it follows that Hill's stability condition in the sense of (4.1) means convexity of the function $\tilde{h}(\log J)$ in $\log J$ or convexity of $\tilde{h}(\operatorname{tr} \log \mathbf{V})$ in $\log \mathbf{V}$.

4.2 Some Volumetric Functions

In the literature, one can find some families of functions $h(J)$, which are used, in particular, in hyperelasticity models for slightly compressible materials. Consider the one-parameter (with parameter $q \in \mathbb{R}$, $q \geq 0$) family of volumetric functions proposed by Hartmann and Neff [1]:

$$h^{(q)}(J) \equiv \begin{cases} \frac{1}{2q^2}(J^q + J^{-q} - 2) = \frac{1}{2}[\frac{1}{q}(J^{q/2} - J^{-q/2})]^2, & \text{if } q > 0, \\ \frac{1}{2}(\ln J)^2 \; [= \lim_{q \to 0} \frac{1}{2q^2}(J^q + J^{-q} - 2)], & \text{if } q = 0. \end{cases} \quad (4.2)$$

Note the symmetry of functions of this family:

$$h^{(q)}(J^{-1}) = h^{(q)}(J).$$

The functions $h^{(q)'}(J)$ for this family have the form

$$h^{(q)'}(J) \; (\equiv \frac{d\, h^{(q)}(J)}{dJ}) = \begin{cases} \frac{1}{2q}(J^{q-1} - J^{-q-1}), & \text{if } q > 0, \\ \ln J / J, & \text{if } q = 0, \end{cases}$$

the functions $J h^{(q)'}(J)$ can be written as

$$J h^{(q)'}(J) = \begin{cases} \frac{1}{2q}(J^q - J^{-q}), & \text{if } q > 0, \\ \ln J, & \text{if } q = 0, \end{cases} \quad (4.3)$$

4.2 Some Volumetric Functions

and the functions $h^{(q)''}(J)$ have the form

$$h^{(q)''}(J) \left(\equiv \frac{d^2 h^{(q)}(J)}{d J^2}\right) = \begin{cases} \frac{1}{2q}[(q-1)J^{q-2} + (q+1)J^{-q-2}], & \text{if } q > 0, \\ J^{-2}(1 - \ln J), & \text{if } q = 0. \end{cases}$$

Other widely used volumetric functions are those from the one-parameter family (with parameter $\beta \in \mathbb{R} \setminus 0$) proposed by Ogden [2]

$$h^{(\beta)}(J) \equiv \beta^{-2}(\beta \ln J + J^{-\beta} - 1). \tag{4.4}$$

The functions $h^{(\beta)'}(J)$ for this family have the form

$$h^{(\beta)'}(J) \left(\equiv \frac{d h^{(\beta)}(J)}{d J}\right) = \beta^{-1}(J^{-1} - J^{-\beta-1}),$$

the function $J h^{(\beta)'}(J)$ can be written as

$$J h^{(\beta)'}(J) = \beta^{-1}(1 - J^{-\beta}),$$

and the functions $h^{(\beta)''}(J)$ have the form

$$h^{(\beta)''}(J) \left(\equiv \frac{d^2 h^{(\beta)}(J)}{d J^2}\right) = \beta^{-1}[(\beta+1)J^{-\beta-2} - J^{-2}].$$

Next, the functions $h^{(q)}(J)$ corresponding to integer values $q = 0, 1, 2, 5$ are identified by ID numbers #1–4, and the functions $h^{(\beta)}(J)$ corresponding to integer values $\beta = -2, -1$ by ID numbers #5 and #6 (see Table 4.1). Volumetric functions #1 and #3–6 with literature references are shown in Table 4 in [1].

Table 4.1 Some volumetric functions $h(J)$ and their properties

ID	Equation	Expression for $h(J)$	Satisfaction of constraints [b]				
			(1)	(2)	(3)[§]	(4)	(5)
1	Eq. (4.2), $q = 0$	$(\ln J)^2/2$	+	+	−	+	+
2	Eq. (4.2), $q = 1$	$(J + J^{-1} - 2)/2$	+	+	+	+	+
3	Eq. (4.2), $q = 2$	$(J^2 + J^{-2} - 2)/8$	+	+	+	+	+
4	Eq. (4.2), $q = 5$	$(J^5 + J^{-5} - 2)/50$	+	+	+	+	+
5	Eq. (4.4), $\beta = -2$	$(J^2 - 2\ln J - 1)/4$	+	+	+	+	+
6	Eq. (4.4), $\beta = -1$	$J - \ln J - 1$	+	+	+	+	+
7	Eq. (4.5)$_1$	$(J-1)^2/2$	+	+	+	−	−
8	Eq. (4.6)$_1$	$(e^{\ln^2 J} - 1)/2$	+	+	+	+	+

[b] +/−, the corresponding constraint is satisfied/not satisfied;
[§] The convexity property (3) does not hold for volumetric function #1 for $J \geq e$ (≈ 2.71)

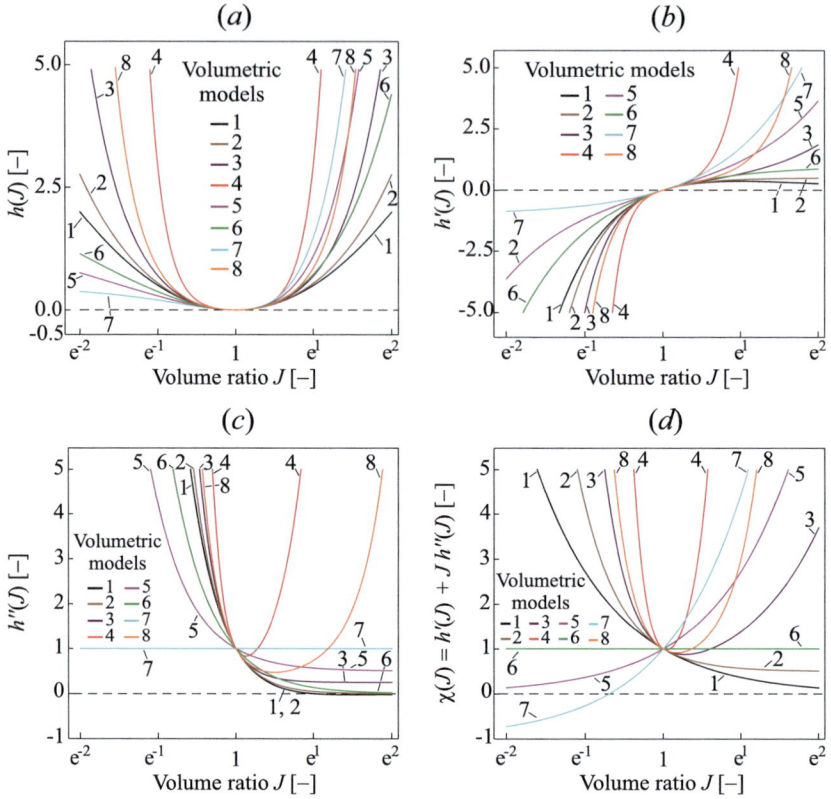

Fig. 4.1 Plots of the functions $h(J)$ **a**, $h'(J)$ **b**, $h''(J)$ **c**, and $\chi(J) = h'(J) + J h''(J)$ **d**

Along with the above volumetric functions, we consider the widely used function (see, e.g., [1], Table 4)

$$h(J) \equiv \frac{1}{2}(J-1)^2 \Rightarrow h'(J) = J-1, \quad J h'(J) = J(J-1), \quad h''(J) = 1. \tag{4.5}$$

This function is denoted by ID number #7 (see Table 4.1).

In addition, we consider the volumetric function

$$h(J) \equiv \frac{1}{2}(e^{\ln^2 J} - 1) \Rightarrow h'(J) = \frac{1}{J} e^{\ln^2 J} \ln J, \quad J h'(J) = e^{\ln^2 J} \ln J, \tag{4.6}$$

$$h''(J) = \frac{1}{J^2}[e^{\ln^2 J}(1 - \ln J + 2\ln^2 J)].$$

This new function is denoted by ID number #8 (see Table 4.1).

Plots of the functions $h(J)$, $h'(J)$, $h''(J)$, and $\chi(J) = h'(J) + J h''(J)$ are presented in Fig. 4.1. These plots provide an answer to the question of whether the

4.2 Some Volumetric Functions

volumetric functions considered here satisfy constraints (1)–(5); these answers are given in Table 4.1. Analysis of the performance of the volumetric functions presented in Table 4.1 shows that not all of these functions satisfy constraints (1)–(5). However, for slightly compressible materials ($J \approx 1$), the plots in Fig. 4.1 demonstrate that the approximate equalities

$$h(J) \approx \frac{1}{2}(\ln J)^2 \approx \frac{1}{2}(J-1)^2, \qquad h'(J) \approx Jh'(J) \approx \ln J \approx J-1$$

are valid in the vicinity of the value $J = 1$.

It is interesting to compare the dependencies of the mean stress $\sigma_m = 1/3(\sigma_1 + \sigma_2 + \sigma_3)$ (or the pressure $p \equiv -\sigma_m$) on the stretch ratio $k (= \lambda_1 = \lambda_2 = \lambda_3)$ for dilatational deformation of the form

$$\mathbf{V} = k\,\mathbf{I} \quad (0 < k < \infty)$$

obtained for the two types of models considered here (mixed and vol-iso) using the volumetric functions presented in Table 4.1 and to compare these dependencies with the experimental data obtained by Bridgman for the volume change of sodium at high pressures [4–7]. Setting $\mu = 2.53$ GPa and $\nu = 0.34$ [8], we obtain the following values for the quantities λ and K:

$$\lambda = \frac{2\mu\nu}{1-2\nu} = 5.37 \text{ GPa}, \qquad K = \lambda + \frac{2}{3}\mu = 7.06 \text{ GPa}.$$

Using the expression $J = k^3$, from (3.12) we obtain the dependence $\sigma_m(k)$ for the mixed models

$$\sigma_m(k) = \frac{\mu}{k^3}(k^2 - 1) + \lambda\,h'(k^3), \qquad (4.7)$$

and from (3.15), the dependence $\sigma_m(k)$ for the vol-iso models

$$\sigma_m(k) = K h'(k^3). \qquad (4.8)$$

Hereinafter, the ID numbers of the mixed and vol-iso models are identified with the ID numbers of volumetric functions (see Tables 1.1, 1.2, and 4.1).[1] Plots of the functions $\sigma_m(k)$ in (4.7) and (4.8) for the mixed and vol-iso models are presented in Fig. 4.2a, and plots of the functions $p(k) = -\sigma_m(k)$ for these models are presented in Fig. 4.2b. Markers in the figures represent the experimental dependencies $\sigma(k)$ and $p(k)$ obtained by Bridgman [4–7]. It can be seen that in the compression region ($k < 1$), the plots of $\sigma_m(k)$ (or $p(k)$) for models #1–6 and #8 are in qualitative agreement with the experimental data of Bridgman (and for models #1-4,8 they are in quantitative agreement, which is consistent with the conclusions in [1]). However,

---

[1] Note that our mixed model #1 is the *Simo–Pister* hyperelastic compressible material model (cf., [9]) and our mixed model #5 is the *Ciarlet–Geymonat* hyperelastic compressible material model (cf., [10]).

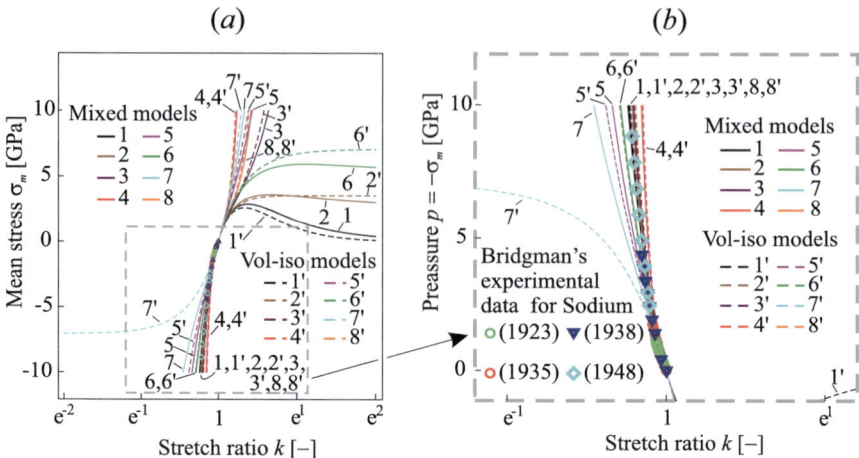

Fig. 4.2 Plots of the functions $\sigma_m(k)$ in (4.7) and (4.8) **a** and plots of the functions $p(k)$ **b**; markers correspond to the experimental values of Bridgman for Sodium

the vol-iso model #7 yields a physically unreasonable response in the form of a finite limiting value for σ_m for $k \to 0$, whereas the mixed model #7 does not lead to a physically unreasonable response. In the extension region ($k > 1$), experimental data on the dependence $\sigma_m(k)$ are difficult to obtain. From the point of view of the idealized concept of deformation, the reasonable response manifested in the tendency of $\sigma_m(k)$ to infinity for $k \to \infty$ is predicted by models #3–8. However, it is models #1 and #2 that correctly describe the response of real materials manifested as loosening under triaxial extension. Obviously, volumetric function #1 provides the best approximation of the dependencies $\sigma_m(k)$ ($k > 1$) for real materials.

References

1. Hartmann, S., Neff, P.: Polyconvexity of generalized polynomial-type hyperelastic strain energy functions for near-incompressibility. Int. J. Solids Struct. **40**(11), 2767–2791 (2003). https://doi.org/10.1016/S0020-7683(03)00086-6
2. Ogden, R.W.: Large deformation isotropic elasticity: on the correlation of theory and experiment for compressible rubberlike solids. Proceedings of the Royal Society of London. A. Math. Phys. Sci. **328**(1575), 567–583 (1972). https://doi.org/10.1098/rspa.1972.0096
3. Neff, P., Graban, K., Schweickert, E., Martin, R.J.: The axiomatic introduction of arbitrary strain tensors by Hans Richter–a commented translation of "Strain tensor, strain deviator and stress tensor for finite deformations." Math. Mech. Solids **25**(5), 1060–1080 (2020). https://doi.org/10.1177/1081286519880594
4. Bridgman, P.W.: The compressibility of thirty metals as a function of pressure and temperature. Proceed. American Acad. Arts Sci. **58**(5), 165–242 (1923)
5. Bridgman, P.W.: Electrical resistances and volume changes up to 20,000 kg/cm^2. Proceed. Nat. Acad. Sci. United States of America **21**(2), 109–113 (1935)

References

6. Bridgman, P.W.: Rough compressibilities of fourteen substances to 45,000 kg/cm^2. Proceed. American Acad. Arts Sci. **72**(5), 207–225 (1938)
7. Bridgman, P.W.: The compression of sixty-one solid substances to 25,000 kg/cm^2, determined by a new rapid method. Proceed. American Acad. Arts Sci. **76**(1), 9–24 (1945)
8. Kellermann, D.C., Attard, M.M.: An invariant-free formulation of neo-Hookean hyperelasticity. Zeitschrift für Angewandte Mathematik und Mechanik **96**(2), 233–252 (2016). https://doi.org/10.1002/zamm.201400210
9. Simo, J.C., Pister, K.S.: Remarks on rate constitutive equations for finite deformation problems: computational implications. Comput. Methods Appl. Mech. Eng. **46**(2), 201–215 (1984). https://doi.org/10.1016/0045-7825(84)90062-8
10. Ciarlet, P.G., Geymonat, G.: Sur les lois de comportement en élasticité non linéaire compressible. Comptes Rendus des Seaces de L'Aademie des Scieces. Serie 2: Mecanique, Physique, Chimie **295**, 423–426 (1982)

Chapter 5
Constitutive Inequalities for Neo-Hookean Materials

Abstract As noted in Chap. 1, in the literature there are some constitutive inequalities that allow the selection of material models with desired properties and/or the establishment of constraints on the parameters included in the constitutive relations for the material models considered. In this book, we test neo-Hookean material models for satisfaction of the constitutive inequalities due to the *Hill postulate* and the *corotational stability postulate* (CSP), which extend the constitutive inequality due to *Drucker's postulate* from infinitesimal to finite strains. Formulations of these inequalities are given in Sect. 5.1, and testing material models for satisfaction of the Hill and CSP postulates is performed in Sects. 5.3 and 5.4, respectively.

5.1 Hill and CSP Constitutive Inequalities

Definition 5.1 (*Drucker's postulate*) (see, e.g., [1–3]) A material model satisfies the *Drucker postulate* (*Drucker's material stability condition*) if under infinitesimal strain conditions for the stress rates $\dot{\sigma}$ satisfying the constitutive relations of this model, the following Drucker's inequality holds at each time instant:

$$\left\langle \frac{\mathrm{D}}{\mathrm{D}t}[\sigma], \frac{\mathrm{D}}{\mathrm{D}t}[\varepsilon] \right\rangle = \dot{\sigma} : \dot{\varepsilon} > 0 \quad \forall \dot{\varepsilon} \neq \mathbf{0}, \quad (5.1)$$

where $\frac{\mathrm{D}}{\mathrm{D}t}$ denotes the material time derivative.

Definition 5.2 (*Hill's postulate*) [4–6] A material model satisfies the *Hill postulate* if for the objective stress rates $\frac{\mathrm{D}^{ZJ}}{\mathrm{D}t}[\tau]$ or $\frac{\mathrm{D}^{BH}}{\mathrm{D}t}[\sigma]$ satisfying the constitutive relations of this model, the following Hill's inequality holds at each time instant:

$$\langle \frac{\mathrm{D}^{ZJ}}{\mathrm{D}t}[\tau], d \rangle = \frac{\mathrm{D}^{ZJ}}{\mathrm{D}t}[\tau] : \mathbf{d} > 0 \quad \Leftrightarrow \quad \langle \frac{\mathrm{D}^{BH}}{\mathrm{D}t}[\sigma], d \rangle = \frac{\mathrm{D}^{BH}}{\mathrm{D}t}[\sigma] : \mathbf{d} > 0 \quad \forall \mathbf{d} \neq \mathbf{0}.$$

© The Author(s), under exclusive license to Springer Nature Switzerland AG 2026
S. Korobeynikov et al., *Two Types of Compressible Isotropic Neo-Hookean Material Models*, SpringerBriefs in Continuum Mechanics,
https://doi.org/10.1007/978-3-032-06050-1_5

Satisfaction of any material model to the Hill postulate is equivalent to the assertion that the *Kirchhoff* stress tensor τ is a monotone function in the left Hencky (logarithmic) strain tensor $\log \mathbf{V} \equiv \sum_{i=1}^{m} \log \lambda_i \mathbf{V}_i$ in the sense that [3] (see also [7])

$$\langle \tau(\log \mathbf{V}_1) - \tau(\log \mathbf{V}_2), \log \mathbf{V}_1 - \log \mathbf{V}_2 \rangle$$
$$= (\tau(\log \mathbf{V}_1) - \tau(\log \mathbf{V}_2)) : (\log \mathbf{V}_1 - \log \mathbf{V}_2) > 0 \quad \forall \mathbf{V}_1 \neq \mathbf{V}_2.$$

In addition, for a hyperelastic solid, the Hill inequality can be equivalently expressed as the convexity of the elastic energy in terms of $\log \mathbf{V}$ since we have the Richter formula [7]

$$\tau = \frac{\partial \widehat{W}(\log \mathbf{V})}{\partial \log \mathbf{V}}, \qquad \widehat{W}(\log \mathbf{V}) := W(\mathbf{V}).$$

Remark 5.3 The equivalence of the formulations of Hill's inequality for $\frac{D^{ZJ}}{Dt}[\tau]$ and $\frac{D^{BH}}{Dt}[\sigma]$ is based on equality (2.26) ($J > 0$).

Definition 5.4 (*Corotational stability postulate*) [1–3, 8] A material model satisfies the *corotational stability postulate* (CSP) if for the corotational stress rates $\frac{D^o}{Dt}[\sigma]$ satisfying the constitutive relations of this model, the following CSP inequality holds at each time instant:

$$\langle \frac{D^o}{Dt}[\sigma], d \rangle = \frac{D^o}{Dt}[\sigma] : \mathbf{d} > 0 \quad \forall \mathbf{d} \neq \mathbf{0}, \tag{5.2}$$

where $\frac{D^o}{Dt}[\sigma]$ is *any* corotational stress rate [9] based on a spin tensor from the family of *continuous material spin tensors* [10–12] (see also [13]). In particular, we can use $\frac{D^{ZJ}}{Dt}[\sigma]$ as this tensor rate.

Satisfaction of any material model to the CSP postulate is equivalent to the assertion that the *Cauchy* stress tensor σ is a monotonic function in the left Hencky (logarithmic) strain tensor $\log \mathbf{V}$ (cf., [3]) meaning that

$$\langle \sigma(\log \mathbf{V}_1) - \sigma(\log \mathbf{V}_2), \log \mathbf{V}_1 - \log \mathbf{V}_2 \rangle$$
$$= (\sigma(\log \mathbf{V}_1) - \sigma(\log \mathbf{V}_2)) : (\log \mathbf{V}_1 - \log \mathbf{V}_2) > 0 \quad \forall \mathbf{V}_1 \neq \mathbf{V}_2.$$

The latter is the **True Stress True Strain Hilbert-Monotonicity** condition (postulate) TSTS-M$^+$ (cf., [3]).

Remark 5.5 Both compressible neo-Hookean formulations (mixed and vol-iso) are polyconvex, therefore rank-one convex provided that μ, K, $\lambda > 0$ and h is convex in J (cf., [14]).

Remark 5.6 The nonequivalence of the constitutive inequalities due to Definitions 5.2 and 5.4 follows from the equality

$$\langle \frac{D^{BH}}{Dt}[\sigma], d \rangle = \frac{D^{BH}}{Dt}[\sigma] : \mathbf{d} = \frac{D^{ZJ}}{Dt}[\sigma] : \mathbf{d} + (\sigma : \mathbf{d}) \operatorname{tr} \mathbf{d}, \quad (5.3)$$

which is a consequence of equality $(2.27)_1$. However, for incompressible materials, $\operatorname{tr} \mathbf{d} = 0$; then it follows from (5.3) that Definitions 5.2 and 5.4 are equivalent for these materials.

Remark 5.7 For infinitesimal strains, both postulates (the Hill and CSP) reduce to the Drucker postulate.

Our present investigation is partly motivated by the fact that polyconvexity or the Legendre–Hadamard ellipticity condition above is clearly not sufficient to guarantee a physically admissible response (see our discussion of the appearance of non-monotonic Cauchy stresses for polyconvex compressible neo-Hookean type models in Chap. 6).

5.2 Testing Linear Elastic Isotropic Material Models Using Drucker's Postulate—Convexity of the Energy

We obtain from $(2.30)_2$ taking into account equality (2.29) the following expression for the contraction for the incompressible material

$$\dot{\sigma} : \dot{\varepsilon} = 2\mu \, \dot{\varepsilon} : \dot{\varepsilon} = 2\mu \, \|\dot{\varepsilon}\|^2 = 2\mu \, \|\operatorname{dev} \dot{\varepsilon}\|^2,$$

from which follows a necessary and sufficient condition for the fulfillment of the inequality of Drucker's postulate (5.1): $\mu > 0$.

Similarly, we obtain from $(2.34)_2$ or $(2.37)_2$ the contraction expression for a compressible material

$$\dot{\sigma} : \dot{\varepsilon} = 2\mu \, \|\operatorname{dev} \dot{\varepsilon}\|^2 + K \, (\operatorname{tr} \dot{\varepsilon})^2. \quad (5.4)$$

The first term on the r.h.s. of (5.4) corresponds to the isochoric strain rate, and the second term corresponds to the volumetric strain rate. Since they cannot be equal to zero simultaneously for an arbitrary non-zero strain rate, the necessary and sufficient conditions for the fulfillment of the Drucker postulate inequality (5.1) are the constraints $\mu, K > 0$ on the parameters of the linear elastic material.

Satisfaction of any linear elastic isotropic material model to the Drucker postulate is equivalent to the assertion that the *Cauchy* stress tensor σ is a monotonic function in the infinitesimal strain tensor ε meaning that

$$\left\langle \frac{D}{Dt}[\sigma], \frac{D}{Dt}[\varepsilon] \right\rangle = \dot{\sigma} : \dot{\varepsilon} > 0 \quad \forall \dot{\varepsilon} \neq 0$$

$\Leftrightarrow \langle \sigma(\varepsilon_1) - \sigma(\varepsilon_2), \varepsilon_1 - \varepsilon_2 \rangle = (\sigma(\varepsilon_1) - \sigma(\varepsilon_2)) : (\varepsilon_1 - \varepsilon_2) \quad \forall \varepsilon_1 \neq \varepsilon_2$

$\Leftrightarrow \varepsilon \to W_{\text{lin}}(\varepsilon)$ is strictly convex

$\Leftrightarrow \mu, K > 0$.

5.3 Testing Neo-Hookean Models Using the Hill Postulate

We test the incompressible neo-Hookean model and the compressible mixed and vol-iso models for satisfaction of the Hill and CSP postulates in Sects. 5.3.1, 5.3.2, and 5.3.3, respectively.

5.3.1 Testing the Incompressible Neo-Hookean Material Model

From (3.4) we obtain the non-objective rate form of constitutive relations for the incompressible neo-Hookean model:

$$\dot{\sigma} = \mu \dot{\mathbf{c}} - \dot{p}\mathbf{I}. \tag{5.5}$$

The equality (see, e.g., [15], p. 126)

$$\frac{D^{\overline{\text{Old}}}}{Dt}[\mathbf{c}] = \mathbf{0}, \quad \frac{D^{\overline{\text{Old}}}}{Dt}[\mathbf{c}] \equiv \dot{\mathbf{c}} - \boldsymbol{\ell} \cdot \mathbf{c} - \mathbf{c} \cdot \boldsymbol{\ell}^T,$$

and Eq. $(2.19)_1$ lead to the following expression for $\dot{\mathbf{c}}$:

$$\dot{\mathbf{c}} = (\mathbf{d} + \mathbf{w}) \cdot \mathbf{c} + \mathbf{c} \cdot (\mathbf{d} - \mathbf{w}). \tag{5.6}$$

The objective rate form of constitutive relations for this model is obtained using $(2.25)_1$, (3.4), (5.5), and (5.6):

$$\frac{D^{\text{ZJ}}}{Dt}[\sigma] = -\dot{p}\mathbf{I} + \mu(\mathbf{d} \cdot \mathbf{c} + \mathbf{c} \cdot \mathbf{d}). \tag{5.7}$$

5.3 Testing Neo-Hookean Models Using the Hill Postulate

Representing the tensor **c** in the form (2.14) and the tensor **d** in the form (2.20) and using the eigenprojection property $(2.4)_1$, we obtain

$$\mathbf{d}\cdot\mathbf{c}+\mathbf{c}\cdot\mathbf{d} = 2\sum_{i=1}^{m}\dot{\lambda}_i\lambda_i\mathbf{V}_i + \sum_{i\neq j=1}^{m}(\lambda_i^2+\lambda_j^2)\mathbf{V}_i\cdot\mathbf{d}\cdot\mathbf{V}_j. \tag{5.8}$$

The contraction is given by

$$A(\mathbf{d}) \equiv (\mathbf{d}\cdot\mathbf{c}+\mathbf{c}\cdot\mathbf{d}) : \mathbf{d}. \tag{5.9}$$

Since the first summand on the r.h.s. of (5.8) and the component $\hat{\mathbf{d}}$ of the tensor **d** in (2.20) are coaxial with the tensor **V** and since the second summand on the r.h.s. of (5.8) and the component $\tilde{\mathbf{d}}$ of the tensor **d** in (2.20) are orthogonal to the tensor **V**, using the eigenprojection property $(2.4)_3$ and (5.9) we obtain

$$A(\mathbf{d}) = 2\sum_{i=1}^{m}(\dot{\lambda}_i)^2 m_i + [\sum_{i\neq j=1}^{m}(\lambda_i^2+\lambda_j^2)\mathbf{V}_i\cdot\mathbf{d}\cdot\mathbf{V}_j] : \tilde{\mathbf{d}}.$$

The quadratic form $A(\mathbf{d})$ can be represented as a sum of two quadratic forms

$$A(\mathbf{d}) = P(\hat{\mathbf{d}}) + R(\tilde{\mathbf{d}}), \tag{5.10}$$

where

$$P(\hat{\mathbf{d}}) \equiv 2\sum_{i=1}^{m}(\dot{\lambda}_i)^2 m_i, \quad R(\tilde{\mathbf{d}}) \equiv \tilde{\mathbf{d}} : \mathbb{R}(\mathbf{V}) : \tilde{\mathbf{d}}, \quad \mathbb{R}(\mathbf{V}) \equiv \sum_{i\neq j=1}^{m}(\lambda_i^2+\lambda_j^2)\mathbf{V}_i \overset{\text{sym}}{\otimes} \mathbf{V}_j. \tag{5.11}$$

According to the statements of Proposition 2.1, the tensor $\mathbb{R}(\mathbf{V}) \in \mathcal{T}^4_{\text{Ssym}}$, and the quadratic form $R(\tilde{\mathbf{d}})$ is a positive definite form of $\tilde{\mathbf{d}}$. Since $P(\hat{\mathbf{d}})$ is also a positive definite form of $\hat{\mathbf{d}}$, it follows that $A(\mathbf{d})$ is a positive definite quadratic form of **d**.

Since

$$\mathbf{I} : \mathbf{d} = \text{tr}\,\mathbf{d} \tag{5.12}$$

and since equality (2.21) is valid, the equality $\text{tr}\,\mathbf{d} = 0$ is valid for incompressible materials ($\dot{J} = 0$). Then from (5.7) we obtain

$$\frac{D^{ZJ}}{Dt}[\sigma] : \mathbf{d} = \mu(\mathbf{d}\cdot\mathbf{c}+\mathbf{c}\cdot\mathbf{d}) : \mathbf{d} = \mu A(\mathbf{d}).$$

Since $A(\mathbf{d})$ is a positive definite quadratic form of $\mathbf{d}$ and $\mu > 0$, we conclude that the incompressible neo-Hookean material model satisfies the Hill postulate.

Remark 5.8 The above conclusion is consistent with the conclusions that the Ogden incompressible material model satisfies the Hill postulate (cf., [16]) and that the incompressible neo-Hookean material model is Ogden's model with one power term to the power $n = 2$.

Remark 5.9 Note that the positive definiteness of the quadratic form $A(\mathbf{d})$ can be proved more simply in the basis-free form for the tensors $\mathbf{c}$ and $\mathbf{d}$, without using the eigenprojections of the tensor $\mathbf{V}$ or, equivalently, the tensor $\mathbf{c}$. We obtain an alternative representation for the quadratic form $A(\mathbf{d})$, based on expressions (5.9)

$$A(\mathbf{d}) = (\mathbf{c} \cdot \mathbf{d}) : \mathbf{d} + (\mathbf{d} \cdot \mathbf{c}) : \mathbf{d} = (\mathbf{c} \cdot \mathbf{d}) : \mathbf{d} + (\mathbf{c}^T \cdot \mathbf{d}^T) : \mathbf{d} = 2(\mathbf{c} \cdot \mathbf{d}) : \mathbf{d}$$
$$= 2(\mathbf{F} \cdot \mathbf{F}^T \cdot \mathbf{d}) : \mathbf{d} = 2(\mathbf{F}^T \cdot \mathbf{d}) : (\mathbf{F}^T \cdot \mathbf{d}) = 2\|\mathbf{F}^T \cdot \mathbf{d}\|^2,$$

from which the positive definiteness of the quadratic form $A(\mathbf{d})$ follows. Nevertheless, the explicit representation of the quadratic form (5.10) turns out to be useful for testing the compressible vol-iso material models (see Sect. 5.3.3).

Alternatively, for the incompressible case, we can write

$$W_{\text{neo-Hooke}}(\mathbf{F}) = \frac{\mu}{2}(\|\mathbf{F}\|^2 - 3) = \frac{\mu}{2}(\|\mathbf{V}\|^2 - 3) = \frac{\mu}{2}(\|\exp(\log \mathbf{V})\|^2 - 3) =: \widehat{W}(\log \mathbf{V}),$$

and it is clear that $\log \mathbf{V} \to \widehat{W}(\log \mathbf{V})$ is convex in $\log \mathbf{V}$, hence Hill's inequality is easily seen to be satisfied.

5.3.2 Testing the Compressible Mixed Material Models

By analogy with the derivation of Eq. (5.7), expression (3.10) for the Kirchhoff stress tensor τ leads to the following objective rate form of constitutive relations for any mixed material model (here we also use equality (2.21)):

$$\frac{D^{ZJ}}{Dt}[\tau] = \mu(\mathbf{d} \cdot \mathbf{c} + \mathbf{c} \cdot \mathbf{d}) + \lambda \chi(J) J \operatorname{tr} \mathbf{d} \, \mathbf{I}, \tag{5.13}$$

the quantity $\chi(J)$ is defined in (4.1). Next, using notation (5.9) and equality (5.12), we get

$$\frac{D^{ZJ}}{Dt}[\tau] : \mathbf{d} = \mu A(\mathbf{d}) + \lambda \chi(J) J (\operatorname{tr} \mathbf{d})^2. \tag{5.14}$$

Note that for $\chi(J) > 0$, the second summand on the r.h.s. of (5.14) is a non-positive definite quadratic form of $\mathbf{d}$. We allow, first, that for this material model, the Lamé

5.3 Testing Neo-Hookean Models Using the Hill Postulate   47

parameter $\lambda = 0$ (and hence $\nu = 0$) (see (3.19)), and, second, that the scalar $\text{tr}\,\mathbf{d}$ can take zero value for $\mathbf{d} \neq 0$, whence it follows that this quadratic form is non-positive definite. However, since $A(\mathbf{d})$ is a positive definite form (see Sect. 5.3.1) and since we assume that $\mu > 0$, the r.h.s. of (5.14) is a positive definite quadratic form of $\mathbf{d}$ for $\chi(J) > 0$. Thus, inequality $\chi(J) > 0$ is a sufficient condition for any mixed compressible neo-Hookean model to satisfy the Hill postulate.

Remark 5.10 The above statement agrees with the conclusion that the Ogden compressible material model satisfies the Hill postulate for $\chi(J) > 0$ (cf., [17]).

5.3.3 Testing the Compressible vol-iso Material Models

Expression (3.14) leads to the following non-objective rate form of constitutive relations for the Kirchhoff stress tensor $\boldsymbol{\tau}$:

$$\dot{\boldsymbol{\tau}} = \mu \overline{\text{dev}\,\bar{\mathbf{c}}} + K\chi(J)J\text{tr}\,\mathbf{d}\,\mathbf{I}. \tag{5.15}$$

Using (2.17), from (2.18)$_2$ we obtain the equality

$$\overline{\text{dev}\,\bar{\mathbf{c}}} = -\frac{2}{3}J^{-5/3}\dot{J}(\mathbf{c} - \frac{1}{3}\text{tr}\,\mathbf{c}\,\mathbf{I}) + J^{-2/3}(\dot{\mathbf{c}} - \frac{1}{3}\text{tr}\,\dot{\mathbf{c}}\,\mathbf{I}). \tag{5.16}$$

In view of equalities (2.21) and (5.6), equality (5.16) can be transformed to

$$\overline{\text{dev}\,\bar{\mathbf{c}}} = J^{-2/3}\left[-\frac{2}{3}(\mathbf{c} - \frac{1}{3}\text{tr}\,\mathbf{c}\,\mathbf{I})\text{tr}\,\mathbf{d} + (\mathbf{d} + \mathbf{w})\cdot\mathbf{c} + \mathbf{c}\cdot(\mathbf{d} - \mathbf{w}) - \frac{1}{3}\text{tr}\,\dot{\mathbf{c}}\,\mathbf{I}\right]. \tag{5.17}$$

Representation (5.6) of the tensor $\dot{\mathbf{c}}$ leads to the equality

$$\text{tr}\,\dot{\mathbf{c}} = \text{tr}(\mathbf{d}\cdot\mathbf{c} + \mathbf{w}\cdot\mathbf{c} + \mathbf{c}\cdot\mathbf{d} - \mathbf{c}\cdot\mathbf{w}) = 2\mathbf{c}:\mathbf{d}, \tag{5.18}$$

which was derived using the equalities

$$\text{tr}(\mathbf{d}\cdot\mathbf{c}) = \text{tr}(\mathbf{c}\cdot\mathbf{d}) = \mathbf{c}:\mathbf{d}, \quad \text{tr}(\mathbf{w}\cdot\mathbf{c}) = \mathbf{w}:\mathbf{c} = 0, \quad \text{tr}(\mathbf{c}\cdot\mathbf{w}) = \mathbf{c}:\mathbf{w} = 0.$$

Using (2.21), (5.17), and (5.18), constitutive relations (3.14), and the definition of the Zaremba–Jaumann stress rate (2.24)$_1$, we transform (5.15) to the objective rate form of constitutive relations for any vol-iso neo-Hookean material model:

$$\frac{D^{ZJ}}{Dt}[\boldsymbol{\tau}] = K\chi(J)J\text{tr}\,\mathbf{d}\,\mathbf{I} + \mu J^{-2/3}\left[-\frac{2}{3}(\mathbf{c} - \frac{1}{3}\text{tr}\,\mathbf{c}\,\mathbf{I})\text{tr}\,\mathbf{d} + \mathbf{d}\cdot\mathbf{c} + \mathbf{c}\cdot\mathbf{d} - \frac{2}{3}(\mathbf{c}:\mathbf{d})\mathbf{I}\right]. \tag{5.19}$$

Performing the contraction of the left- and right-hand sides of Eq. (5.19) with the tensor **d** and using equality (5.12), we arrive at the equality

$$\frac{D^{ZJ}}{Dt}[\boldsymbol{\tau}]:\mathbf{d} = K\chi(J)J(\operatorname{tr}\mathbf{d})^2 + \mu J^{-2/3}[-\frac{4}{3}(\mathbf{c}:\mathbf{d})\operatorname{tr}\mathbf{d} + \frac{2}{9}\operatorname{tr}\mathbf{c}(\operatorname{tr}\mathbf{d})^2 + (\mathbf{d}\cdot\mathbf{c}+\mathbf{c}\cdot\mathbf{d}):\mathbf{d}]. \tag{5.20}$$

Since the unit tensor **I** is coaxial with the tensor **V**, using the property $(2.4)_3$ of eigenprojections and equalities (2.14) and (2.20), we obtain

$$\operatorname{tr}\mathbf{d} = \mathbf{d}:\mathbf{I} = \hat{\mathbf{d}}:\mathbf{I} = \sum_{i=1}^{m}\frac{\dot{\lambda}_i}{\lambda_i}m_i, \quad \operatorname{tr}\mathbf{c} = \sum_{i=1}^{m}\lambda_i^2 m_i, \quad \mathbf{c}:\mathbf{d} = \mathbf{c}:\hat{\mathbf{d}} = \sum_{i=1}^{m}\dot{\lambda}_i\lambda_i m_i. \tag{5.21}$$

We introduce the quadratic forms

$$B(\hat{\mathbf{d}}) \equiv (\mathbf{c}:\hat{\mathbf{d}})\operatorname{tr}\hat{\mathbf{d}}, \quad C(\hat{\mathbf{d}}) \equiv \operatorname{tr}\mathbf{c}\,(\operatorname{tr}\hat{\mathbf{d}})^2. \tag{5.22}$$

In view of expressions (5.21), the quadratic forms (5.22) can be rewritten in explicit form

$$B(\hat{\mathbf{d}}) = \left(\sum_{i=1}^{m}\dot{\lambda}_i\lambda_i m_i\right)\left(\sum_{j=1}^{m}\frac{\dot{\lambda}_j}{\lambda_j}m_j\right), \quad C(\hat{\mathbf{d}}) = \left(\sum_{i=1}^{m}\lambda_i^2 m_i\right)\left(\sum_{j=1}^{m}\frac{\dot{\lambda}_j}{\lambda_j}m_j\right)^2. \tag{5.23}$$

The quadratic form in square brackets on the r.h.s. of (5.20) will be denoted by

$$D(\mathbf{d}) \equiv -\frac{4}{3}B(\hat{\mathbf{d}}) + \frac{2}{9}C(\hat{\mathbf{d}}) + A(\mathbf{d}).$$

According to the representation (5.10) of the quadratic form $A(\mathbf{d})$, the quantity $D(\mathbf{d})$ can be written as

$$D(\mathbf{d}) = E(\hat{\mathbf{d}}) + R(\tilde{\mathbf{d}}), \tag{5.24}$$

where

$$E(\hat{\mathbf{d}}) \equiv -\frac{4}{3}B(\hat{\mathbf{d}}) + \frac{2}{9}C(\hat{\mathbf{d}}) + P(\hat{\mathbf{d}}). \tag{5.25}$$

Our next goal is to determine the properties of the quadratic form $E(\hat{\mathbf{d}})$. Consider the general case $m = 3$, i.e., $m_i = 1$ ($i = 1, 2, 3$). We use expressions $(5.11)_1$ and (5.22) for the quadratic forms on the right-hand side of (5.25). After a series of transformations, we arrive at the final expression for the quadratic form $E(\hat{\mathbf{d}})$:

5.3 Testing Neo-Hookean Models Using the Hill Postulate

$$E(x_1, x_2, x_3) \equiv x_1^2(4 + a^{-2} + b^{-2}) + x_2^2(4 + a^2 + c^{-2}) \qquad (5.26)$$
$$+ x_3^2(4 + b^2 + c^2) + x_1 x_2(-4a - 4a^{-1} + 2b^{-1}c^{-1})$$
$$+ x_1 x_3(-4b - 4b^{-1} + 2a^{-1}c) + x_2 x_3(-4c - 4c^{-1} + 2ab),$$

where the following notations are used:

$$x_1 \equiv \dot{\lambda}_1^2, \quad x_2 \equiv \dot{\lambda}_2^2, \quad x_3 \equiv \dot{\lambda}_3^2, \quad a \equiv \frac{\lambda_1}{\lambda_2}, \quad b \equiv \frac{\lambda_1}{\lambda_3}, \quad c \equiv \frac{\lambda_2}{\lambda_3} \quad (a > 0, \, b > 0, \, c > 0).$$

In view of the identity

$$(x - y - z)^2 = x^2 + y^2 + z^2 - 2xy - 2xz - 2yz,$$

which is valid for all $x, y, z \in \mathbb{R}$, expression (5.26) for the quadratic form $E(x_1, x_2, x_3)$ can be simplified to

$$E(x_1, x_2, x_3) = (2x_1 - x_2 a - x_3 b)^2 + (2x_2 - x_1 a^{-1} - x_3 c)^2 + (2x_3 - x_1 b^{-1} - x_2 c^{-1})^2. \qquad (5.27)$$

It follows from (5.27) that $E(\hat{\mathbf{d}}) \geq 0 \, \forall \, \hat{\mathbf{d}}$ and $\exists \, \hat{\mathbf{d}} \neq \mathbf{0}$ are such that $E(\hat{\mathbf{d}}) = 0$; i.e., the quadratic form $E(\hat{\mathbf{d}})$ is semi-positive definite. Since the quadratic form $R(\tilde{\mathbf{d}})$ is positive definite, it follows from (5.24) that the quantity $D(\mathbf{d})$ is semi-positive definite, i.e.,

$$D(\mathbf{d}) \geq 0 \, \forall \, \mathbf{d} \quad \text{and} \quad \exists \, \mathbf{d} \neq \mathbf{0} \text{ that } D(\mathbf{d}) = 0.$$

The expression for the quadratic form $E(x_1, x_2, x_3)$ for $m = 2$ is derived from expression (5.27) by setting in the latter

$$x_1 \equiv \dot{\lambda}_1^2, \quad x_2 = x_3 \equiv \dot{\lambda}_2^2, \quad a = b \equiv \frac{\lambda_1}{\lambda_2}, \quad c = 1 \quad (a = b > 0).$$

For $m = 1$, we get the identities

$$x_1 = x_2 = x_3 \equiv \dot{\lambda}_1^2, \quad a = b = c = 1,$$

whence it follows that $E(\hat{\mathbf{d}}) = 0 \, \forall \, \hat{\mathbf{d}}$. Since in this case, $R(\tilde{\mathbf{d}}) = 0$, we have $D(\mathbf{d}) = 0 \, \forall \, \mathbf{d}$; i.e., only the first summand is retained on the right-hand side of (5.20).

Returning to expression (5.20), we note that for $\chi(J) > 0$, the first summand on the r.h.s. is a semi-positive definite quadratic form which can vanish for body motions with $\operatorname{tr} \mathbf{d} = 0$ for $\mathbf{d} \neq \mathbf{0}$. Since the second summand on the r.h.s. is also a semi-positive definite quadratic form, we need to determine whether both forms can vanish simultaneously when $\mathbf{d} \neq \mathbf{0}$.

An explicit expression for the contraction $\operatorname{tr} \mathbf{d}$ is presented in $(5.21)_1$. We now determine conditions under which the quadratic form (5.27) vanishes for nonzero values of x_1, x_2, and x_3. Since in this case, each expression in parentheses on the r.h.s. of (5.27) must vanish, we come to the search for nontrivial solutions of the system of homogeneous equation

$$\mathbf{A}\mathbf{x} = \tilde{\mathbf{0}}, \quad \mathbf{A} \equiv \begin{bmatrix} 2 & -a & -b \\ -a^{-1} & 2 & -c \\ -b^{-1} & -c^{-1} & 2 \end{bmatrix}, \quad \mathbf{x} \equiv \begin{bmatrix} x_1 \\ x_2 \\ x_3 \end{bmatrix}, \quad \tilde{\mathbf{0}} \equiv \begin{bmatrix} 0 \\ 0 \\ 0 \end{bmatrix}. \quad (5.28)$$

We are interested in nontrivial solutions of system (5.28) that are possible when the quantity $\det \mathbf{A}$ vanishes, whence we obtain

$$\det \mathbf{A} = (b - ac)^2 \Rightarrow \det \mathbf{A} = 0 \Rightarrow b = ac \Leftrightarrow \frac{\lambda_1}{\lambda_3} = \frac{\lambda_1}{\lambda_3};$$

i.e., the equality $\det \mathbf{A} = 0$ holds for any values of the principal stretches λ_1, λ_2, and λ_3. Thus, for any values of the principal stretches, there exist values of the quantities $\dot{\lambda}_1$, $\dot{\lambda}_2$, and $\dot{\lambda}_3$, not equal to zero simultaneously, for which the quadratic form $E(\dot{\lambda}_1, \dot{\lambda}_2, \dot{\lambda}_3)$ vanishes. The only nontrivial solution of the system is

$$x_1 \in R, \ x_2 = a^{-1}x_1, \ x_3 = b^{-1}x_1 \Leftrightarrow \frac{\dot{\lambda}_1}{\lambda_1} \in \mathbb{R}, \ \frac{\dot{\lambda}_2}{\lambda_2} = \frac{\dot{\lambda}_1}{\lambda_1}, \ \frac{\dot{\lambda}_3}{\lambda_3} = \frac{\dot{\lambda}_1}{\lambda_1}. \quad (5.29)$$

A particular case of this solution is the solution for $m = 1$ ($\lambda_1 = \lambda_2 = \lambda_3, \dot{\lambda}_1 = \dot{\lambda}_2 = \dot{\lambda}_3$) considered above (we established that in this case, $E(\hat{\mathbf{d}}) = 0$). Comparing the expressions in (5.29) and (5.21)$_1$, we see that the quantities $(\operatorname{tr}\mathbf{d})^2$ and $E(\hat{\mathbf{d}})$ cannot vanish simultaneously for motions with $\hat{\mathbf{d}} \neq \mathbf{0}$. The results of our analysis lead to the conclusion that any vol-iso neo-Hookean material model with $\chi(J) > 0$ satisfies the Hill postulate.

An alternative derivation of Hill's inequality is based on the convexity property of elastic energy for the vol-iso material models. Let us represent the elastic energy for these material models in the form ($V = \sqrt{F F^T}$)

$$W_{\text{vol-iso}}(F) = \frac{\mu}{2} \left\| \frac{F}{(\det F)^{\frac{1}{3}}} \right\|^2 + K\, h(\det F) = \frac{\mu}{2} \left\| \frac{V}{(\det V)^{\frac{1}{3}}} \right\|^2 + K\, h(\det V) =: \widehat{W}(\log V). \quad (5.30)$$

We shall show that $\log V \mapsto \widehat{W}(\log V)$ is strictly convex if $h: \mathbb{R}^+ \to \mathbb{R}$ satisfies $h''(e^\xi)\, e^\xi + h'(e^\xi) > 0$ for all $\xi \in \mathbb{R}$. First, we have to express $\widehat{W}$. Using the properties of $\log V$ it holds that

$$\log \det V = \operatorname{tr}(\log V) \iff \det V = e^{\operatorname{tr}(\log V)}, \quad (5.31)$$

$$\frac{V}{(\det V)^{\frac{1}{3}}} = \frac{\exp(\log V)}{(\det V)^{\frac{1}{3}}} = \frac{\exp(\log V)}{e^{\frac{1}{3}\operatorname{tr}(\log V)}} = \exp(\log V - \frac{1}{3}\operatorname{tr}(\log V)\, I) = \exp(\operatorname{dev} \log V). \quad (5.32)$$

5.4 Testing Neo-Hookean Models Using the CSP

Therefore,

$$\widehat{W}(\log V) = \frac{\mu}{2} \| \exp(\text{dev} \log V) \|^2 + K\, h(e^{\text{tr}(\log V)}),$$
$$\widehat{W}(S) = \frac{\mu}{2} \| \exp(\text{dev}\, S) \|^2 + K\, h(e^{\text{tr}(S)}). \tag{5.33}$$

Note that $\text{tr}(S) \mapsto e^{\text{tr}(S)}$ is clearly convex, and if $h\colon \mathbb{R}^+ \to \mathbb{R}$ satisfies $h''(e^\xi) e^\xi + h'(e^\xi) > 0$ for all $\xi \in \mathbb{R}$, then the composition $\text{tr}(S) \mapsto h(e^{\text{tr}(S)})$ is strictly convex in $\text{tr}(S)$. This amounts to $\chi(J) > 0$. Moreover, it is well known that $X \mapsto \| \exp(X) \|^2$ is strictly convex [5, 18], therefore $\text{dev}\, S \mapsto \| \exp(\text{dev}\, S) \|^2$ is strictly convex. Since $S = \text{dev}\, S + \frac{1}{3} \text{tr}(S) \mathbf{I}$ this shows that for $\mu, K > 0$ the function $S \mapsto \widehat{W}(S)$ is strictly convex. Using Richter's formula [19–23] for the Kirchhoff stress

$$\tau = \text{D}_{\log V} \widehat{W}(\log V) \tag{5.34}$$

we observe that τ must then be strictly monotone in $\log V$, i.e.

$$\langle \tau(\log V_1) - \tau(\log V_2), \log V_1 - \log V_2 \rangle > 0 \quad \forall V_1 \neq V_2 \tag{5.35}$$

and this is equivalent to Hill's inequality.

5.4 Testing Neo-Hookean Models Using the CSP

As noted in Remark 5.6, the Hill postulate and the CSP are equivalent for incompressible materials, and since we established in Sect. 5.3.1 that the incompressible neo-Hookean model satisfies the Hill postulate, this model also satisfies the CSP.

We now need to determine whether any mixed model satisfies inequality (5.2). In view of (2.26)$_1$, Eq. (5.3) leads to the following expression for the contraction on the left-hand side of inequality (5.2):

$$\frac{\text{D}^{\text{ZJ}}}{\text{D}t}[\sigma] : \mathbf{d} = \frac{1}{J} \frac{\text{D}^{\text{ZJ}}}{\text{D}t}[\tau] : \mathbf{d} - (\sigma : \mathbf{d})\, \text{tr}\, \mathbf{d}. \tag{5.36}$$

Using expression (5.14) for the contraction $\frac{\text{D}^{\text{ZJ}}}{\text{D}t}[\tau] : \mathbf{d}$ and expression (3.12) for the Cauchy stress tensor σ, from (5.36) we obtain

$$\frac{\text{D}^{\text{ZJ}}}{\text{D}t}[\sigma] : \mathbf{d} = \lambda J h''(J)(\text{tr}\, \mathbf{d})^2 + \frac{\mu}{J}[A(\mathbf{d}) + (\text{tr}\, \mathbf{d})^2 - (\mathbf{c} : \mathbf{d})\, \text{tr}\, \mathbf{d}], \tag{5.37}$$

where the quadratic form $A(\mathbf{d})$ is defined in (5.9). We introduce the quadratic form

$$F(\hat{\mathbf{d}}) \equiv (\operatorname{tr}\hat{\mathbf{d}})^2 (= (\operatorname{tr}\mathbf{d})^2) = \left(\sum_{i=1}^{m} \frac{\dot{\lambda}_i}{\lambda_i} m_i\right)^2, \tag{5.38}$$

where equality $(5.21)_1$ is used. In view of equalities (5.10), $(5.22)_1$, and (5.38), the expression on the r.h.s. of (5.37) can be rewritten as

$$\frac{\mathrm{D}^{ZJ}}{\mathrm{D}t}[\boldsymbol{\sigma}] : \mathbf{d} = \lambda J h''(J) F(\hat{\mathbf{d}}) + \frac{\mu}{J}[P(\hat{\mathbf{d}}) + R(\tilde{\mathbf{d}}) + F(\hat{\mathbf{d}}) - B(\hat{\mathbf{d}})].$$

We are interested in the properties of the quadratic form

$$G(\hat{\mathbf{d}}) \equiv P(\hat{\mathbf{d}}) + F(\hat{\mathbf{d}}) - B(\hat{\mathbf{d}}).$$

Equalities $(5.11)_1$, $(5.23)_1$, and (5.38) lead to the explicit expression for this quadratic form

$$G(\hat{\mathbf{d}}) = 2\sum_{i=1}^{m}(\dot{\lambda}_i)^2 m_i + \left(\sum_{i=1}^{m}\frac{\dot{\lambda}_i}{\lambda_i}m_i\right)^2 - \left(\sum_{i=1}^{m}\dot{\lambda}_i\lambda_i m_i\right)\left(\sum_{j=1}^{m}\frac{\dot{\lambda}_j}{\lambda_j}m_j\right).$$

This quadratic form is neither positive nor semi-positive definite. This statement can be illustrated by the following example. Let $m = 1$, i.e., let $\lambda_1 = \lambda_2 = \lambda_3$ and $m_1 = 3$. Then

$$G(\hat{\mathbf{d}}) = 3(\dot{\lambda}_1)^2\left(\frac{3}{\lambda_1^2} - 1\right),$$

and for $\lambda_1 > \sqrt{3}$, we have $G(\hat{\mathbf{d}}) < 0$. Thus, the quadratic form $\frac{\mathrm{D}^{ZJ}}{\mathrm{D}t}[\boldsymbol{\sigma}] : \mathbf{d}$ is generally neither positive definite nor even semi-positive definite. Hence any mixed neo-Hookean model does not satisfy the CSP.

The fact that the vol-iso neo-Hookean models do not satisfy the CSP has been shown by Neff et al. (cf., [3], Appendix A.4).

Remark 5.11 While no model of our neo-Hookean material compressible family does satisfy the CSP $\Leftrightarrow$ TSTS-M$^+$ condition, it is known that the so-called exponential Hencky energy [3, 24]

$$W_{\text{exp-Hencky}}(\mathbf{F}) = \mu \left(e^{\|\log \mathbf{V}\|^2} - 1\right) + \frac{\lambda}{2}\left(e^{\log^2 J} - 1\right) \tag{5.39}$$

satisfies TSTS-M$^+$ for $\mu, \lambda > 0$. However, (5.39) is not polyconvex or LH-elliptic throughout. The last author has offered a 500 euro challenge for finding a compressible isotropic energy that satisfies simultaneously TSTS-M$^+$ and polyconvexity [25].

References

1. d'Agostino, M.V., Holthausen, S., Bernardini, D., Sky, A., Ghiba, I.D., Martin, R.J., Neff, P.: A constitutive condition for idealized isotropic Cauchy elasticity involving the logarithmic strain. J. Elast. **157**(1), 23 (2025). https://doi.org/10.1007/s10659-024-10097-2
2. Neff, P., Husemann, N.J., Nguetcho Tchakoutio, A.S., Korobeynikov, S.N., Martin, R.J.: The corotational stability postulate: Positive incremental Cauchy stress moduli for diagonal, homogeneous deformations in isotropic nonlinear elasticity. Int. J. Non-Linear Mech. **174**, 105033 (2025). https://doi.org/10.1016/j.ijnonlinmec.2025.105033
3. Neff, P., Holthausen, S., d'Agostino, M.V., Bernardini, D., Sky, A., Ghiba, I.D., Martin, R.J.: Hypo-elasticity, Cauchy-elasticity, corotational stability and monotonicity in the logarithmic strain. J. Mech. Phys. Solids **202**, 106074 (2025). https://doi.org/10.1016/j.jmps.2025.106074
4. Hill, R.: On constitutive inequalities for simple materials-I. J. Mech. Phys. Solids **16**(4), 229–242 (1968). https://doi.org/10.1016/0022-5096(68)90031-8
5. Hill, R.: Constitutive inequalities for isotropic elastic solids under finite strain. Proceedings of the Royal Society of London. A. Math. Phys. Sci. **314**(1519), 457–472 (1970). https://royalsocietypublishing.org/doi/abs/10.1098/rspa.1970.0018
6. Hill, R.: Aspects of invariance in solid mechanics. In: Yih, C.S. (ed.) Advances in Applied Mechanics, vol. 18, pp. 1–75. Academic Press, New York (1979). https://doi.org/10.1016/S0065-2156(08)70264-3
7. Neff, P., Graban, K., Schweickert, E., Martin, R.J.: The axiomatic introduction of arbitrary strain tensors by Hans Richter-a commented translation of "Strain tensor, strain deviator and stress tensor for finite deformations." Math. Mech. Solids **25**(5), 1060–1080 (2020). https://doi.org/10.1177/1081286519880594
8. Ghiba, I.D., Martin, R.J., Apetrii, M., Neff, P.: Constitutive properties for isotropic energies in ideal nonlinear elasticity for solid materials: numerical evidences for invertibility and monotonicity in different stress-strain pairs (in preparation)
9. Neff, P., Holthausen, S., Korobeynikov, S.N., Ghiba, I.D., Martin, R.J.: A natural requirement for objective corotational rates-on structure-preserving corotational rates. Acta Mech. **236**(4), 2657–2689 (2025). https://doi.org/10.1007/s00707-025-04249-1
10. Xiao, H., Bruhns, O., Meyers, A.: On objective corotational rates and their defining spin tensors. Int. J. Solids Struct. **35**(30), 4001–4014 (1998). https://doi.org/10.1016/S0020-7683(97)00267-9
11. Xiao, H., Bruhns, O.T., Meyers, A.: Strain rates and material spins. J. Elast. **52**(1), 1–41 (1998). https://doi.org/10.1023/A:1007570827614
12. Xiao, H., Bruhns, O., Meyers, A.: Objective corotational rates and unified work-conjugacy relation between Eulerian and Lagrangean strain and stress measures. Arch. Mech. **50**(6), 1015–1045 (1998)
13. Korobeynikov, S.N.: Families of continuous spin tensors and applications in continuum mechanics. Acta Mech. **216**(1–4), 301–332 (2011). https://doi.org/10.1007/s00707-010-0369-7
14. Hartmann, S., Neff, P.: Polyconvexity of generalized polynomial-type hyperelastic strain energy functions for near-incompressibility. Int. J. Solids Struct. **40**(11), 2767–2791 (2003). https://doi.org/10.1016/S0020-7683(03)00086-6
15. Korobeynikov, S.N.: Objective tensor rates and applications in formulation of hyperelastic relations. J. Elast. **93**(2), 105–140 (2008). https://doi.org/10.1007/s10659-008-9166-0

16. Ogden, R.W.: Large deformation isotropic elasticity—on the correlation of theory and experiment for incompressible rubberlike solids.Proceedings of the Royal Society of London. A. Mathematical and Physical Sciences **326**(1567), 565–584 (1972). https://doi.org/10.1098/rspa.1972.0026
17. Ogden, R.W.: Large deformation isotropic elasticity: on the correlation of theory and experiment for compressible rubberlike solids. Proceedings of the Royal Society of London. A. Mathematical and Physical Sciences **328**(1575), 567–583 (1972). https://doi.org/10.1098/rspa.1972.0096
18. Ogden, R.W.: Non-linear Elastic Deformations. Ellis Horwood, Chichester (1984)
19. Richter, H.: Das isotrope Elastizitätsgesetz. Z. Angew. Math. Mech. **28**(7–8), 205–209 (1948). https://doi.org/10.1002/zamm.19480280703
20. Richter, H.: Hauptaufsätze: Verzerrungstensor, Verzerrungsdeviator und Spannungstensor bei endlichen Formänderungen. Z. Angew. Math. Mech. **29**(3), 65–75 (1949). https://doi.org/10.1002/zamm.19490290301
21. Richter, H.: Zum Logarithmus einer Matrix. Arch. Math. **2**(5), 360–363 (1949). https://doi.org/10.1007/BF02036865
22. Richter, H.: Über Matrixfunktionen. Math. Ann. **122**(1), 16–34 (1950). https://doi.org/10.1007/BF01342947
23. Richter, H.: Zur Elastizitätstheorie endlicher Verformungen. Mathematische Nachrichten **8**(1), 65–73 (1952). https://doi.org/10.1002/mana.19520080109
24. Neff, P., Ghiba, I.D., Lankeit, J.: The exponentiated Hencky-logarithmic strain energy. Part I: Constitutive issues and rank-one convexity. J. Elast. **121**(2), 143–234 (2015). https://doi.org/10.1007/s10659-015-9524-7
25. Neff, P.: Truesdell's Hauptproblem: On constitutive stability in idealized isotropic nonlinear elasticity and a 500 euro challenge **10.13140/RG.2.2.15349.69603**, (Presentation) (2025)

Chapter 6
Testing Neo-Hookean Materials in Homogeneous Deformations

Abstract In this chapter, general forms of expressions for stress and strain/deformation tensors for all types of homogeneous deformations considered here are presented in Sect. 6.1. The dependencies of kinematic and static quantities on prescribed stretches for uniaxial loading, equibiaxial loading in plane stress, and uniaxial loading in plain strain are given in Sects. 6.2, 6.3, and 6.4, respectively.

6.1 Forms of Stress and Strain/Deformation Tensors for Homogeneous Deformations

Following Kossa et al. [1], we consider three types of homogeneous deformation (Fig. 6.1): *uniaxial loading (UL)* (Fig. 6.1a), *equibiaxial loading in plane stress (ELP)* (Fig. 6.1b), and *uniaxial loading in plane strain (ULP)* (Fig. 6.1c). In these three cases, the tensors $\mathbf{F}$, $\mathbf{c}$, $\boldsymbol{\sigma}$, and $\mathbf{P}$ have the following forms:

$$\mathbf{F} = \begin{bmatrix} \lambda_1 & 0 & 0 \\ 0 & \lambda_2 & 0 \\ 0 & 0 & \lambda_3 \end{bmatrix}, \quad \mathbf{c} = \begin{bmatrix} \lambda_1^2 & 0 & 0 \\ 0 & \lambda_2^2 & 0 \\ 0 & 0 & \lambda_3^2 \end{bmatrix}, \quad (6.1)$$

$$\boldsymbol{\sigma} = \begin{bmatrix} \sigma_{11} & 0 & 0 \\ 0 & \sigma_{22} & 0 \\ 0 & 0 & \sigma_{33} \end{bmatrix}, \quad \mathbf{P} = \begin{bmatrix} P_{11} & 0 & 0 \\ 0 & P_{22} & 0 \\ 0 & 0 & P_{33} \end{bmatrix},$$

and the volume ratio J has the form (2.15). The Finger strain tensor $\mathbf{e}^{(2)}$ is written as follows (see (2.13)):

$$\mathbf{e}^{(2)} = \frac{1}{2}\begin{bmatrix} \lambda_1^2 - 1 & 0 & 0 \\ 0 & \lambda_2^2 - 1 & 0 \\ 0 & 0 & \lambda_3^2 - 1 \end{bmatrix}. \quad (6.2)$$

Using Eqs. (2.17) and (2.18)$_2$, we obtain the following expression for the tensor dev $\bar{\mathbf{c}}$:

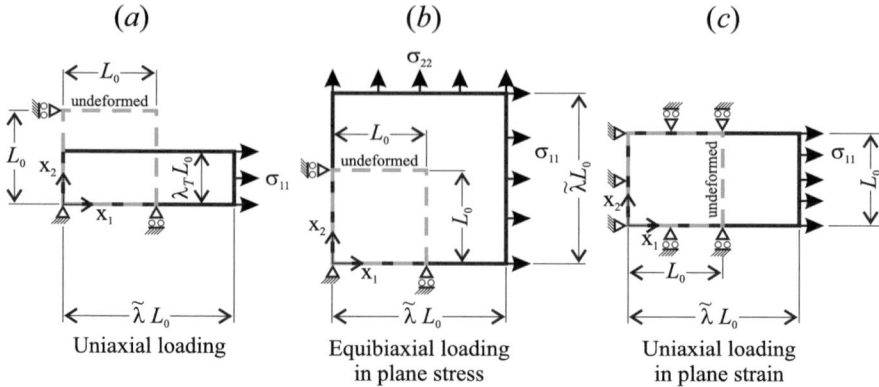

Fig. 6.1 Sketches of homogeneous deformations: uniaxial loading **a**, equibiaxial loading in plane stress **b**, and uniaxial loading in plane strain **c**

$$\mathrm{dev}\,\bar{\mathbf{c}} = \frac{1}{3} J^{-2/3} \begin{bmatrix} 2\lambda_1^2 - \lambda_2^2 - \lambda_3^2 & 0 & 0 \\ 0 & -\lambda_1^2 + 2\lambda_2^2 - \lambda_3^2 & 0 \\ 0 & 0 & -\lambda_1^2 - \lambda_2^2 + 2\lambda_3^2 \end{bmatrix}. \quad (6.3)$$

We now specify the expressions for the quantities J, $\mathbf{e}^{(2)}$, and dev $\bar{\mathbf{c}}$ for the homogeneous deformations considered (hereinafter, the stretch $\tilde{\lambda}$ is assumed to be a prescribed quantity).

- For UL (Fig. 6.1a), the stress-strain state is subjected to the following constraints:

$$\lambda_1 = \tilde{\lambda}, \qquad \sigma_{22} = \sigma_{33} = 0, \qquad P_{22} = P_{33} = 0. \quad (6.4)$$

Transverse strains in UL will be denoted by

$$\lambda_T \equiv \lambda_2 = \lambda_3. \quad (6.5)$$

The volume ratio J is determined from (2.15), (6.4)$_1$, and (6.5)

$$J = \tilde{\lambda}\lambda_T^2. \quad (6.6)$$

For UL, the Finger strain tensor $\mathbf{e}^{(2)}$ defined in (6.2) has the form

$$\mathbf{e}^{(2)} = \frac{1}{2}\begin{bmatrix} \tilde{\lambda}^2 - 1 & 0 & 0 \\ 0 & \lambda_T^2 - 1 & 0 \\ 0 & 0 & \lambda_T^2 - 1 \end{bmatrix},$$

and in view of expressions (6.3), (6.4)$_1$, and (6.5), the tensor dev $\bar{\mathbf{c}}$ can be written as

6.1 Forms of Stress and Strain/Deformation Tensors for Homogeneous Deformations

$$\text{dev}\,\bar{\mathbf{c}} = \frac{1}{3}J^{-2/3}\begin{bmatrix} 2(\tilde{\lambda}^2 - \lambda_T^2) & 0 & 0 \\ 0 & \lambda_T^2 - \tilde{\lambda}^2 & 0 \\ 0 & 0 & \lambda_T^2 - \tilde{\lambda}^2 \end{bmatrix}. \qquad (6.7)$$

Using (2.23), (6.1), and (6.4), we obtain the following expression for the quantity P_{11}:

$$P_{11} = \lambda_T^2 \sigma_{11}. \qquad (6.8)$$

- For ELP (Fig. 6.1b), the stress-strain state is subjected to the following constraints:

$$\lambda_1 = \lambda_2 = \tilde{\lambda}, \qquad \sigma_{33} = 0, \qquad \sigma_{11} = \sigma_{22}, \qquad P_{11} = P_{22}, \qquad P_{33} = 0. \qquad (6.9)$$

Transverse strains in ELP will be denoted by

$$\lambda_T \equiv \lambda_3. \qquad (6.10)$$

The volume ratio J is determined from (2.15), (6.9)$_1$, and (6.10):

$$J = \tilde{\lambda}^2 \lambda_T. \qquad (6.11)$$

For ELP, the Finger strain tensor $\mathbf{e}^{(2)}$ defined in (6.2) has the form

$$\mathbf{e}^{(2)} = \frac{1}{2}\begin{bmatrix} \tilde{\lambda}^2 - 1 & 0 & 0 \\ 0 & \tilde{\lambda}^2 - 1 & 0 \\ 0 & 0 & \lambda_T^2 - 1 \end{bmatrix}, \qquad (6.12)$$

and in view of expressions (6.3), (6.9)$_1$, and (6.10), the tensor dev $\bar{\mathbf{c}}$ can be written as

$$\text{dev}\,\bar{\mathbf{c}} = \frac{1}{3}J^{-2/3}\begin{bmatrix} \tilde{\lambda}^2 - \lambda_T^2 & 0 & 0 \\ 0 & \tilde{\lambda}^2 - \lambda_T^2 & 0 \\ 0 & 0 & 2(\lambda_T^2 - \tilde{\lambda}^2) \end{bmatrix}. \qquad (6.13)$$

In view of (2.23), (6.1), and (6.11), the nonzero components of the first P-K stress tensor can be written as

$$P_{11} = P_{22} = \tilde{\lambda}\lambda_T \sigma_{11}. \qquad (6.14)$$

- For ULP (Fig. 6.1c), the stress-strain state is subjected to the following constraints:

$$\lambda_1 = \tilde{\lambda}, \qquad \lambda_2 = 1, \qquad \sigma_{33} = 0, \qquad P_{33} = 0. \qquad (6.15)$$

Transverse strains in ULP will be denoted by

$$\lambda_T \equiv \lambda_3. \qquad (6.16)$$

The volume ratio J is determined from (2.15), (6.15)$_{1,2}$, and (6.16):

$$J = \tilde{\lambda}\lambda_T. \tag{6.17}$$

For ULP, the Finger strain tensor $\mathbf{e}^{(2)}$ defined in (6.2) has the form

$$\mathbf{e}^{(2)} = \frac{1}{2}\begin{bmatrix} \tilde{\lambda}^2 - 1 & 0 & 0 \\ 0 & 0 & 0 \\ 0 & 0 & \lambda_T^2 - 1 \end{bmatrix}, \tag{6.18}$$

and in view of expressions (6.3), (6.15)$_{1,2}$, and (6.16), the tensor dev $\bar{\mathbf{c}}$ can be written as

$$\text{dev}\,\bar{\mathbf{c}} = \frac{1}{3}J^{-2/3}\begin{bmatrix} 2\tilde{\lambda}^2 - 1 - \lambda_T^2 & 0 & 0 \\ 0 & -\tilde{\lambda}^2 + 2 - \lambda_T^2 & 0 \\ 0 & 0 & 2\lambda_T^2 - 1 - \tilde{\lambda}^2 \end{bmatrix}. \tag{6.19}$$

In view of (2.23), (6.1), and (6.17), the nonzero components of the first P-K stress tensor can be written as

$$P_{11} = \lambda_T \sigma_{11}, \quad P_{22} = \tilde{\lambda}\lambda_T \sigma_{22}. \tag{6.20}$$

Kossa et al. [1] used the vol-iso material model with volumetric energy given by the expression $\frac{K}{2}(J^2 - 1)$ (i.e., using volumetric function #7) and varied Poisson's ratio in the range $-1 \leq \nu \leq 0.5$. Since the bulk modulus K is non-negative in this range of Poisson's ratio and since the functions $h(J)$ used in the present work are non-negative (see Sect. 4), it follows that the volumetric energy is also non-negative. For the mixed models, the non-negativity of the volumetric energy $\lambda h(J)$ is determined by the non-negativity of the parameter λ. Since the non-negativity of this parameter is guaranteed by the range of Poisson's ratio $0 \leq \nu \leq 0.5$, in order for the volumetric energies to be non-negative for both material models (see, e.g., [2]), here we restrict ourselves to the following set of values of Poisson's ratio ν for compressible neo-Hookean materials:

$$\nu = \{0, 0.25, 0.4, 0.45, 0.499, 0.4999\}. \tag{6.21}$$

The value $\nu = 0.5$ is assigned to the classical incompressible neo-Hookean model.

6.2 Uniaxial Loading

The dependencies of stresses and unknown lateral principal stretches on the prescribed longitudinal stretch obtained by solving the uniaxial loading problem for the incompressible isotropic neo-Hookean material model and mixed and vol-iso

6.2 Uniaxial Loading

Table 6.1 Limiting values of λ_T, σ_{11}, and P_{11} in extreme states where $\tilde{\lambda} \to 0$ and $\tilde{\lambda} \to \infty$ in the UL and ELP problems for the incompressible isotropic neo-Hookean material model

Quantity	$\tilde{\lambda} \to 0$	$\tilde{\lambda} \to \infty$
$\lambda_T(\tilde{\lambda})$	$+\infty$	0
$\sigma_{11}(\tilde{\lambda})$	$-\infty$	$+\infty$
$P_{11}(\tilde{\lambda})$	$-\infty$	$+\infty$

compressible isotropic neo-Hookean material models are presented in Sects. 6.2.1, 6.2.2, and 6.2.3, respectively.

6.2.1 Incompressible Isotropic Neo-Hookean Material

Setting $J = 1$, from (6.6) we obtain

$$\lambda_T = \tilde{\lambda}^{-1/2}. \tag{6.22}$$

In view of (3.4) and (6.2), the components of the Cauchy stress tensor can be written as

$$\sigma_{11} = \mu\,(\tilde{\lambda}^2 - 1) - p, \qquad \sigma_{22} = \sigma_{33} = \mu\,(\lambda_T^2 - 1) - p. \tag{6.23}$$

Determining the Lagrange multiplier p from (6.4)$_2$ and (6.23)$_2$ and using (6.22), we obtain

$$\sigma_{11} = \mu\,(\tilde{\lambda}^2 - \tilde{\lambda}^{-1}). \tag{6.24}$$

In view of (6.22) and (6.24), from (6.8) we get

$$P_{11} = \tilde{\lambda}^{-1}\sigma_{11} = \mu\,(\tilde{\lambda} - \tilde{\lambda}^{-2}). \tag{6.25}$$

The limiting values of λ_T, σ_{11}, and P_{11} in extreme states are obtained from expressions (6.22), (6.24), and (6.25) and presented in Table 6.1. We assume that these limiting values correspond to physically reasonable responses for idealized hyperelastic materials.

6.2.2 Compressible Isotropic Mixed Neo-Hookean Material Models

In view of (3.12) and (6.2), the components of the Cauchy stress tensor can be written as

$$\sigma_{11} = \lambda\, h'(J) + \frac{\mu}{J}(\tilde{\lambda}^2 - 1), \qquad \sigma_{22} = \sigma_{33} = \lambda\, h'(J) + \frac{\mu}{J}(\lambda_T^2 - 1). \qquad (6.26)$$

Using (6.4)$_2$ and (6.6), we obtain the nonlinear implicit dependence of λ_T on $\tilde{\lambda}$ in the general case:

$$\lambda\, h'(\tilde{\lambda}\lambda_T^2) + \frac{\mu}{\tilde{\lambda}}(1 - \lambda_T^{-2}) = 0. \qquad (6.27)$$

We first consider the value $\nu = 0$ for Poisson's ratio. Since for this value of ν, $\lambda = 0$ (see (3.16)$_2$), from (6.27) we obtain

$$\lambda_T = 1. \qquad (6.28)$$

Then from (6.6), (6.8) and (6.26)$_1$, we get

$$\sigma_{11} = P_{11} = \mu\,(\tilde{\lambda} - \tilde{\lambda}^{-1}). \qquad (6.29)$$

The values of λ_T in (6.28) and σ_{11} and P_{11} in (6.29) are valid for any functions $h'(J)$. The fact that for $\nu = 0$, there is no lateral deformation for any value $\tilde{\lambda}$ is considered a physically reasonable result (see, e.g., [3]) by analogy with deformation under UL when using the equations of linear elasticity theory.

For the remaining values of ν from the interval $0 < \nu < 0.5$, the value of λ_T should be determined from the nonlinear equation (6.27). In the particular case of mixed model #7, the function $h'(J)$ has the form (4.5)$_2$. In this case, Eq. (6.27) has an explicit solution (cf., [4], Eq. (4.25))

$$\lambda_T = \left[\frac{1}{2a}(-b + \sqrt{b^2 - 4ac})\right]^{1/2}, \qquad (6.29a)$$

where

$$a \equiv \lambda\tilde{\lambda}, \qquad b \equiv \frac{\mu}{\tilde{\lambda}}, \qquad c \equiv -\frac{\mu}{\tilde{\lambda}}.$$

For the remaining volumetric functions $h'(J)$ considered in this book, the dependence $\lambda_T(\tilde{\lambda})$ is determined from (6.27) using the Wolfram Mathematica software. Substitution of the obtained dependence into (6.26)$_1$ with the use of expression (6.6) yields the dependence $\sigma_{11}(\tilde{\lambda})$. Using the dependencies $\lambda_T(\tilde{\lambda})$ and $\sigma_{11}(\tilde{\lambda})$, from (6.8) we find the dependence $P_{11}(\tilde{\lambda})$.

Plots of λ_T versus $\tilde{\lambda}$ for mixed models #1–4 are shown in Fig. 6.2 along with dependencies (6.22) for the incompressible neo-Hookean material model ($\nu = 0.5$). From the plots and an analysis of the solutions using the Wolfram Mathematica software, it follows that the limiting values of λ_T for these material models with $\nu \neq 0$ (see Table 6.2) agree with the physically reasonable limiting values in Table 6.1. Note also that the obtained plots for mixed model #1 agree with the plots obtained for the same material model by Ehlers and Eipper (cf., [5], Fig. 3, curve (a)).

6.2 Uniaxial Loading

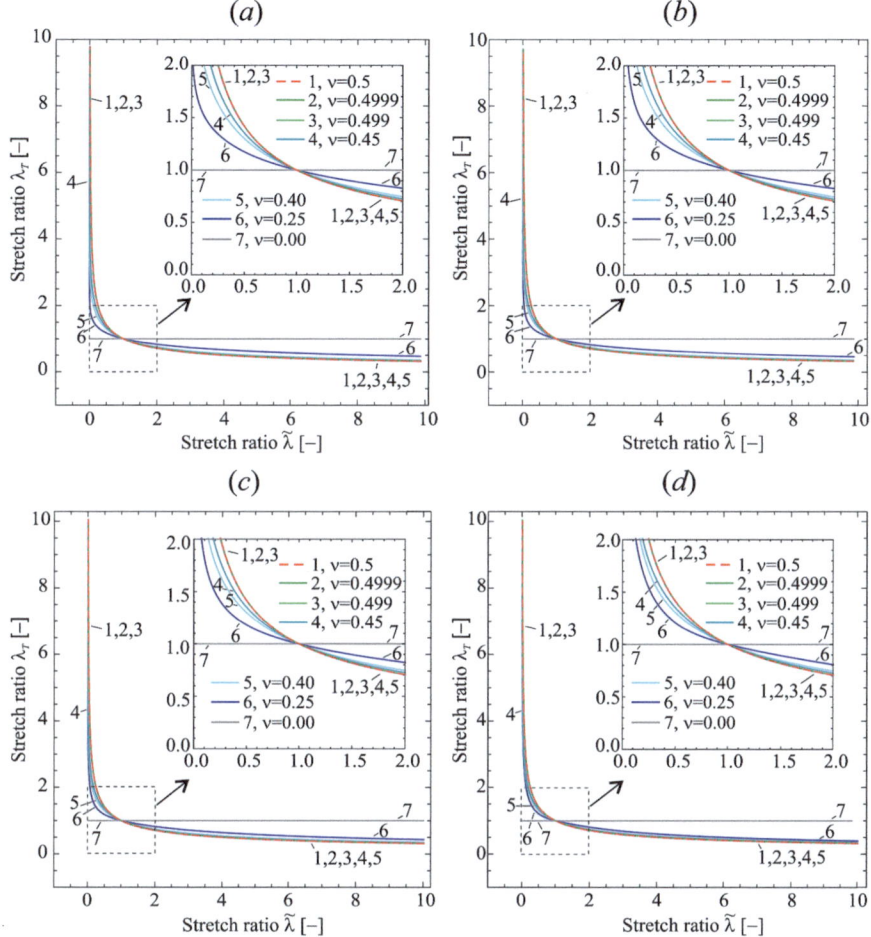

Fig. 6.2 Plots of λ_T versus $\tilde{\lambda}$ in the UL problem for mixed models #1 **a**, #2 **b**, #3 **c**, and #4 **d**

Plots of λ_T versus $\tilde{\lambda}$ for mixed models #5–8 are presented in Fig. 6.3. Analysis of the solutions using the Wolfram Mathematica software shows that for $\lambda_T \to 0$, the limiting values of λ_T for material models #5–7 with $\nu \neq 0$ do not agree with the physically reasonable limiting values in Table 6.1. In particular, using (6.29a), for mixed model #7 we obtain the same value $\lim_{\tilde{\lambda} \to 0} \lambda_T = 1$ for all values of Poisson's ratio $0 < \nu < 0.5$ (see also [4], Eq. (4.26)$_1$). The dependencies $\lambda_T(\tilde{\lambda})$ obtained for model #7 (see Fig. 6.3c) agree with the dependencies derived by Pence and Gou (cf., [4], Fig. 6). For mixed models #5 and #6, different limiting values of λ_T for $\tilde{\lambda} \to 0$ are obtained using different values of Poisson's ratio from the set (6.21) (see Table 6.2).

Table 6.2 Limiting values of λ_T, σ_{11}, and P_{11} in extreme states where $\tilde{\lambda} \to 0$ and $\tilde{\lambda} \to \infty$ in the UL problem for compressible isotropic material models with $0 \le \nu < 0.5$

Model ID	Quantity[b]	Mixed models		Vol-iso models	
		$\tilde{\lambda} \to 0$	$\tilde{\lambda} \to \infty$	$\tilde{\lambda} \to 0$	$\tilde{\lambda} \to \infty$
1	$\lambda_T(\tilde{\lambda})$	$+\infty$	0	**0**	**$+\infty$**
	$\sigma_{11}(\tilde{\lambda})$	$-\infty$	$+\infty$	$-\infty$	**0**
	$P_{11}(\tilde{\lambda})$	$-\infty$	$+\infty$	$-\infty$	**0**
2	$\lambda_T(\tilde{\lambda})$	$+\infty$	0	$+\infty$	$+\infty$
	$\sigma_{11}(\tilde{\lambda})$	$-\infty$	$+\infty$	$-\infty$	*
	$P_{11}(\tilde{\lambda})$	$-\infty$	$+\infty$	$-\infty$	$+\infty$
3	$\lambda_T(\tilde{\lambda})$	$+\infty$	0	$+\infty$	0
	$\sigma_{11}(\tilde{\lambda})$	$-\infty$	$+\infty$	$-\infty$	$+\infty$
	$P_{11}(\tilde{\lambda})$	$-\infty$	$+\infty$	$-\infty$	$+\infty$
4	$\lambda_T(\tilde{\lambda})$	$+\infty$	0	$+\infty$	0
	$\sigma_{11}(\tilde{\lambda})$	$-\infty$	$+\infty$	$-\infty$	$+\infty$
	$P_{11}(\tilde{\lambda})$	$-\infty$	$+\infty$	$-\infty$	$+\infty$
5	$\lambda_T(\tilde{\lambda})$	*	0	**0**	0
	$\sigma_{11}(\tilde{\lambda})$	$-\infty$	$+\infty$	$-\infty$	$+\infty$
	$P_{11}(\tilde{\lambda})$	$-\infty$	$+\infty$	$-\infty$	$+\infty$
6	$\lambda_T(\tilde{\lambda})$	*	0	**0**	**$+\infty$**
	$\sigma_{11}(\tilde{\lambda})$	$-\infty$	$+\infty$	$-\infty$	$+\infty$
	$P_{11}(\tilde{\lambda})$	$-\infty$	$+\infty$	$-\infty$	$+\infty$
7	$\lambda_T(\tilde{\lambda})$	1	0	**0**	0
	$\sigma_{11}(\tilde{\lambda})$	$-\infty$	$+\infty$	**$-3K$**	$+\infty$
	$P_{11}(\tilde{\lambda})$	$-\infty$	$+\infty$	**0**	$+\infty$
8	$\lambda_T(\tilde{\lambda})$	$+\infty$	0	$+\infty$	0
	$\sigma_{11}(\tilde{\lambda})$	$-\infty$	$+\infty$	$-\infty$	$+\infty$
	$P_{11}(\tilde{\lambda})$	$-\infty$	$+\infty$	$-\infty$	$+\infty$

[b] An asterisk (∗) denotes some finite limiting values, and standard and highlighted texts indicate physically reasonable and unreasonable values of a quantity, respectively

Plots of the Cauchy stress σ_{11} versus the stretch $\tilde{\lambda}$ for mixed material models #1–4 and #5–8 are shown in Figs. 6.4 and 6.5, respectively. Similar plots of the engineering (1st P-K, nominal) stress P_{11} versus the stretch $\tilde{\lambda}$ for mixed material models #1–4 and #5–8 are shown in Figs. 6.6 and 6.7, respectively. We see that the stresses obtained for all compressible material models considered in this book under slight compressibility conditions (with $\nu = 0.499$ and $\nu = 0.4999$) are close to the stresses for the incompressible nH model (with $\nu = 0.5$). Note that the plots of $\sigma_{11}(\tilde{\lambda})$ obtained for mixed model #7 (see Fig. 6.5c) agree with the plots obtained by Pence and Gou for the same material model (cf., [4], Fig. 7). We used the Wolfram Mathematica software to determine the limiting values of σ_{11} and P_{11} in extreme states and obtained the limiting values of these quantities (see Table 6.2) that agree with the physically reasonable values presented in Table 6.1.

6.2 Uniaxial Loading

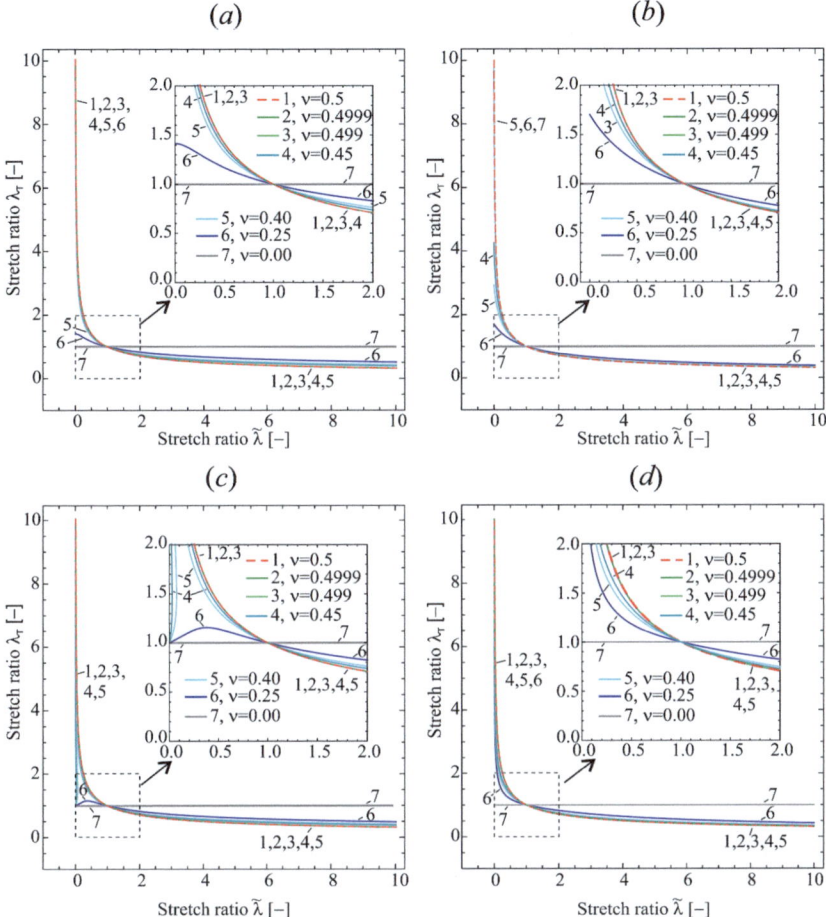

Fig. 6.3 Plots of λ_T versus $\tilde{\lambda}$ in the UL problem for mixed models #5 **a**, #6 **b**, #7 **c**, and #8 **d**

The solution of the UL problem for the mixed models leads to the conclusion that the most reliable physically reasonable solutions can be obtained using volumetric functions from the Hartmann–Neff family and volumetric function #8. First, for values of Poisson's ratio $\nu = 0$, there is no lateral strength. Second, for values of Poisson's ratio $\nu \neq 0$ for material models with these volumetric functions in extreme states, the lateral stretch λ_T and the stresses σ_{11} and P_{11} have physically reasonable limiting values that agree with the limiting values of these quantities for incompressible materials. Third, the solutions for λ_T, σ_{11}, and P_{11} for mixed models with $\nu = 0.499$ and $\nu = 0.4999$ are close to the corresponding solutions for the classical incompressible nH material model.

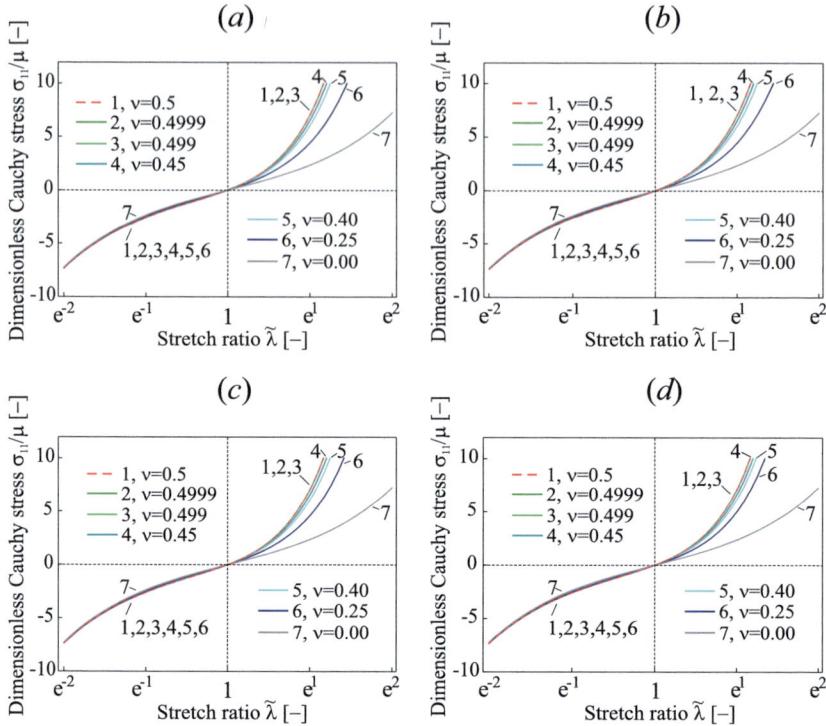

Fig. 6.4 Plots of σ_{11} versus $\tilde{\lambda}$ in the UL problem for mixed models #1 **a**, #2 **b**, #3 **c**, and #4 **d**

6.2.3 Compressible Isotropic vol-iso Neo-Hookean Material Models

In view of (3.15) and (6.7), the components of the Cauchy stress tensor can be written as

$$\sigma_{11} = K\, h'(J) + \frac{2}{3}\mu\, J^{-5/3}(\tilde{\lambda}^2 - \lambda_T^2), \tag{6.30}$$

$$\sigma_{22} = \sigma_{33} = K\, h'(J) + \frac{1}{3}\mu\, J^{-5/3}(\lambda_T^2 - \tilde{\lambda}^2).$$

Regardless of the choice of Poisson's ratio $\nu \in [0, 0.5)$, Eqs. (6.4)$_2$, (6.6), and (6.30)$_2$ lead to the following nonlinear equation for the dependence λ_T vs. $\tilde{\lambda}$:

$$K\, h'(\tilde{\lambda}\lambda_T^2) + \frac{1}{3}\mu\, J^{-5/3}(\lambda_T^2 - \tilde{\lambda}^2) = 0.$$

Summing the left and right sides of equalities in (6.30) and taking into account equalities $\sigma_{22} = \sigma_{33} = 0$, we obtain

6.2 Uniaxial Loading

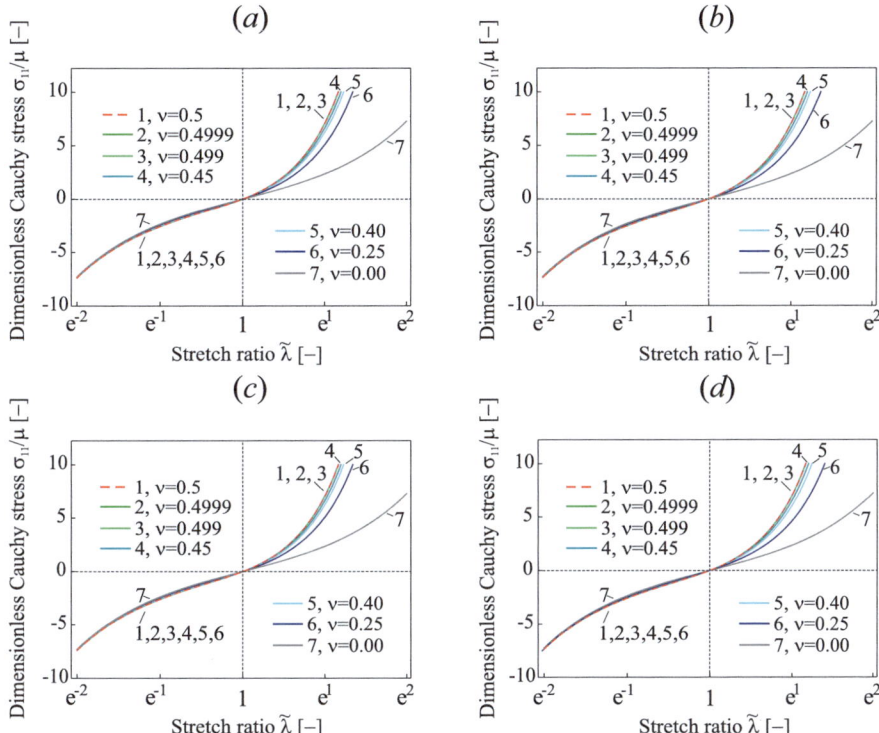

Fig. 6.5 Plots of σ_{11} versus $\tilde{\lambda}$ in the UL problem for mixed models #5 **a**, #6 **b**, #7 **c**, and #8 **d**

$$\sigma_{11} = 3K\, h'(\tilde{\lambda}\lambda_T^2). \tag{6.31}$$

Substitution of $\lambda_T(\tilde{\lambda})$ into the r.h.s. of (6.31) yields the dependence $\sigma_{11}(\tilde{\lambda})$. The dependence $P_{11}(\tilde{\lambda})$ is obtained from (6.8) using the dependencies $\lambda_T(\tilde{\lambda})$ and $\sigma_{11}(\tilde{\lambda})$.

Plots of λ_T versus $\tilde{\lambda}$ for vol-iso models #1–4, and #5–8 are presented in Figs. 6.8 and 6.9, respectively, along with dependencies (6.22) for incompressible nH material ($\nu = 0.5$). Analysis of the solutions using the Wolfram Mathematica software shows that the limiting values of λ_T in extreme states for vol-iso material models with $0 \le \nu < 0.5$ agree with the physically reasonable limiting values in Table 6.1 only for the models #3,4,8 (see Table 6.2). For the remaining models (#1,2,5,6,7), data on the limiting physically unreasonable values of λ_T are presented in Table 6.2. Note that the plots of $\lambda_T(\tilde{\lambda})$ in Fig. 6.8a for vol-iso model #1 agree with the plot (a) in Fig. 2 in [5] for the same material model. The plots for vol-iso model #3 (see Fig. 6.8c) agree with the plots for the same material model in Fig. 9 in [4]. In addition, the plots of $\lambda_T(\tilde{\lambda})$ in Fig. 6.9c for vol-iso material model #7 agree with the plots in Fig. 2 in [1] for the same material model.

Plots of the Cauchy stress σ_{11} versus the stretch $\tilde{\lambda}$ for vol-iso material models #1-4 and #5-8 are shown in Figs. 6.10 and 6.11, respectively. Similar plots of the

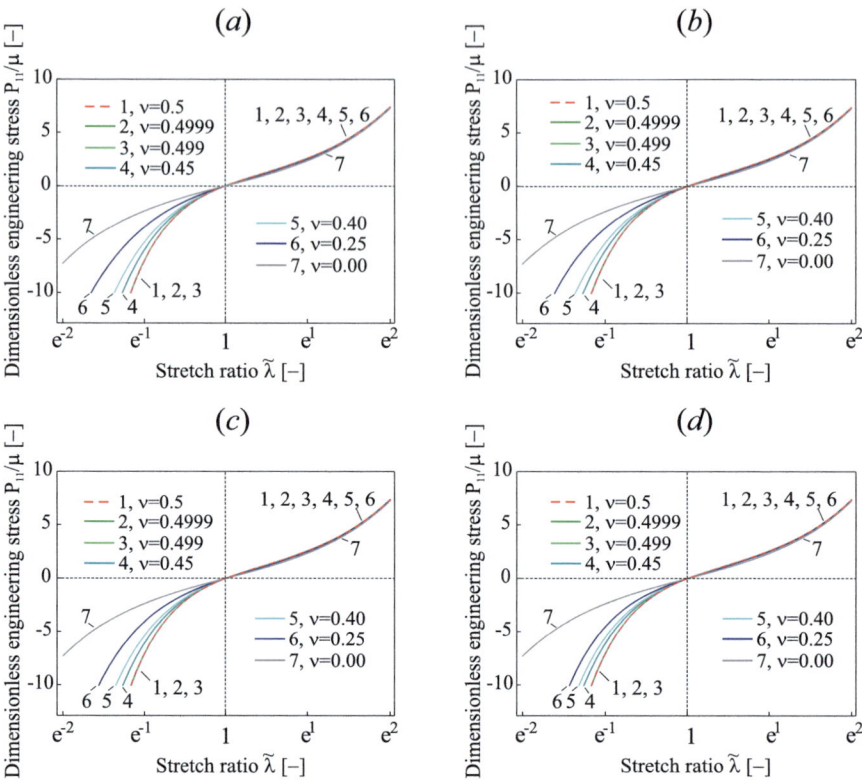

Fig. 6.6 Plots of P_{11} versus $\tilde{\lambda}$ in the UL problem for mixed models #1 **a**, #2 **b**, #3 **c**, and #4 **d**

engineering (1st P-K, nominal) stress P_{11} versus the stretch $\tilde{\lambda}$ for vol-iso material models #1–4 and #5–8 are shown in Figs. 6.12 and 6.13, respectively.[1] Note that the plots of the stress σ_{11} versus the stretch $\tilde{\lambda}$ for vol-iso model #3 in Fig. 6.10c agree with the plots in Fig. 10 in [4] for the same material model (the model of W_b in terms of [4]). In addition, the plots $\sigma_{11}(\tilde{\lambda})$ and $P_{11}(\tilde{\lambda})$ in Figs. 6.11c and 6.13c for vol-iso material model #7 agree with the plots in Figs. 14a and 5 in [1] for the same material model. We observe the non-monotonic dependencies of Cauchy stresses on stretches in Fig. 6.11c for this material model. This is clearly physically inadmissible as long as the elastic material is not damaged.

The limiting values of σ_{11} and P_{11} in extreme states obtained for vol-iso material models with $0 \leq \nu < 0.5$ using the Wolfram Mathematica software are given in Table 6.2. It can be seen that physically reasonable limiting values exist only for vol-iso models #3–6,8 (see Table 6.1).

---

[1] To overcome the difficulties in deriving the dependencies $\sigma_{11}(\tilde{\lambda})$ and $P_{11}(\tilde{\lambda})$ for vol-iso material model #1 using the Wolfram Mathematica software, we approximated the function $\ln J$ by the function $6(J^{1/12} - J^{-1/12})$ (i.e., we used the value $q = 1/12$ in (4.3)).

6.3 Equibiaxial Loading in Plane Stress

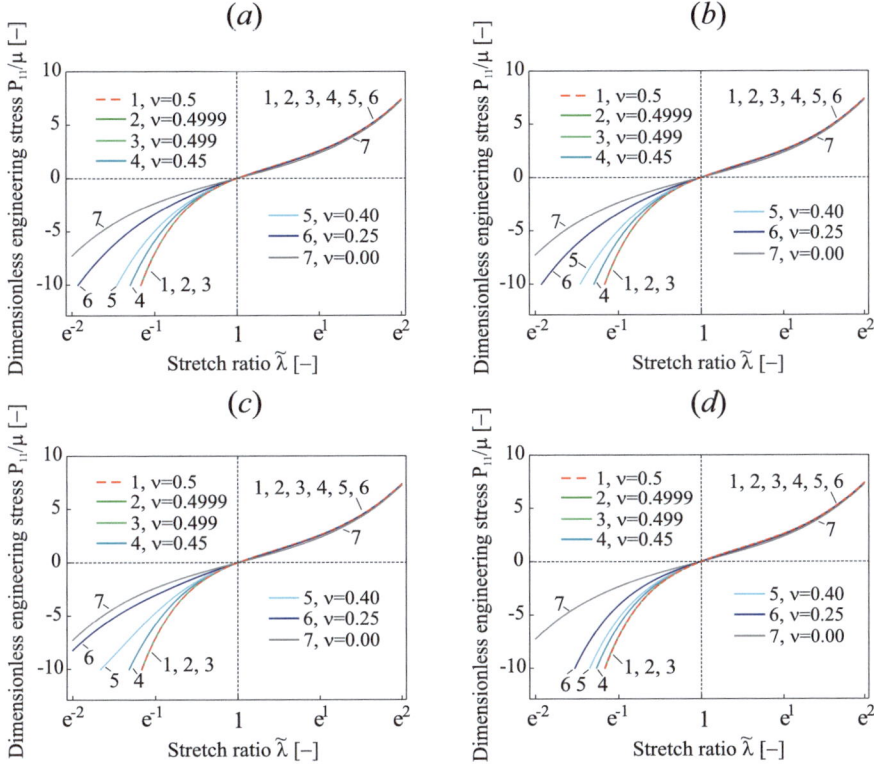

Fig. 6.7 Plots of P_{11} versus $\tilde{\lambda}$ in the UL problem for mixed models #5 **a**, #6 **b**, #7 **c**, and #8 **d**

The results of the solution of the UL problem for vol-iso models lead to the conclusion that physically reasonable solutions can be obtained only using volumetric functions #3,4,8.

6.3 Equibiaxial Loading in Plane Stress

The dependencies of stresses and the unknown out-of-plane principal stretch versus prescribed in-plane principal stretches obtained by solving the problem of equibiaxial loading in plane stress for the incompressible isotropic neo-Hookean material model and compressible mixed and vol-iso neo-Hookean models are presented in Sects. 6.3.1, 6.3.2, and 6.3.3, respectively.

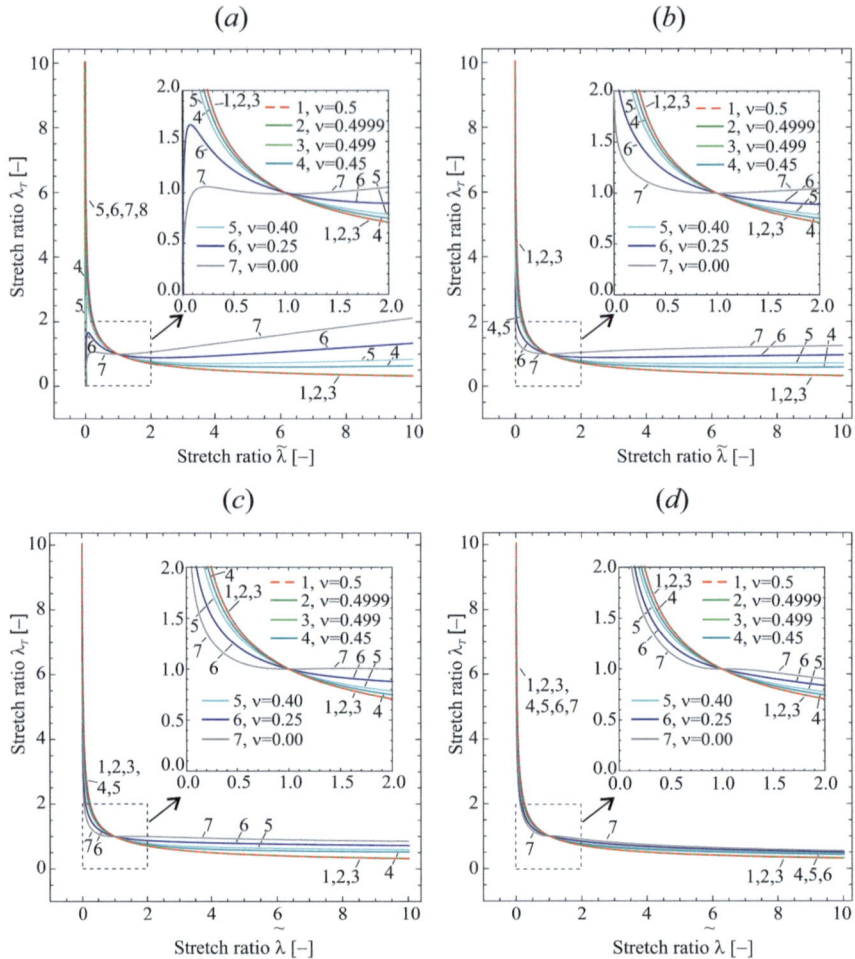

Fig. 6.8 Plots of λ_T versus $\tilde{\lambda}$ in the UL problem for vol-iso models #1 **a**, #2 **b**, #3 **c**, and #4 **d**

6.3.1 Incompressible Isotropic Neo-Hookean Material

Setting $J = 1$, from (6.11) we obtain

$$\lambda_T = \tilde{\lambda}^{-2}. \tag{6.32}$$

In view of (3.4) and (6.12), the components of the Cauchy stress tensor can be written as

$$\sigma_{11} = \sigma_{22} = \mu\,(\tilde{\lambda}^2 - 1) - p, \qquad \sigma_{33} = \mu\,(\lambda_T^2 - 1) - p. \tag{6.33}$$

6.3 Equibiaxial Loading in Plane Stress

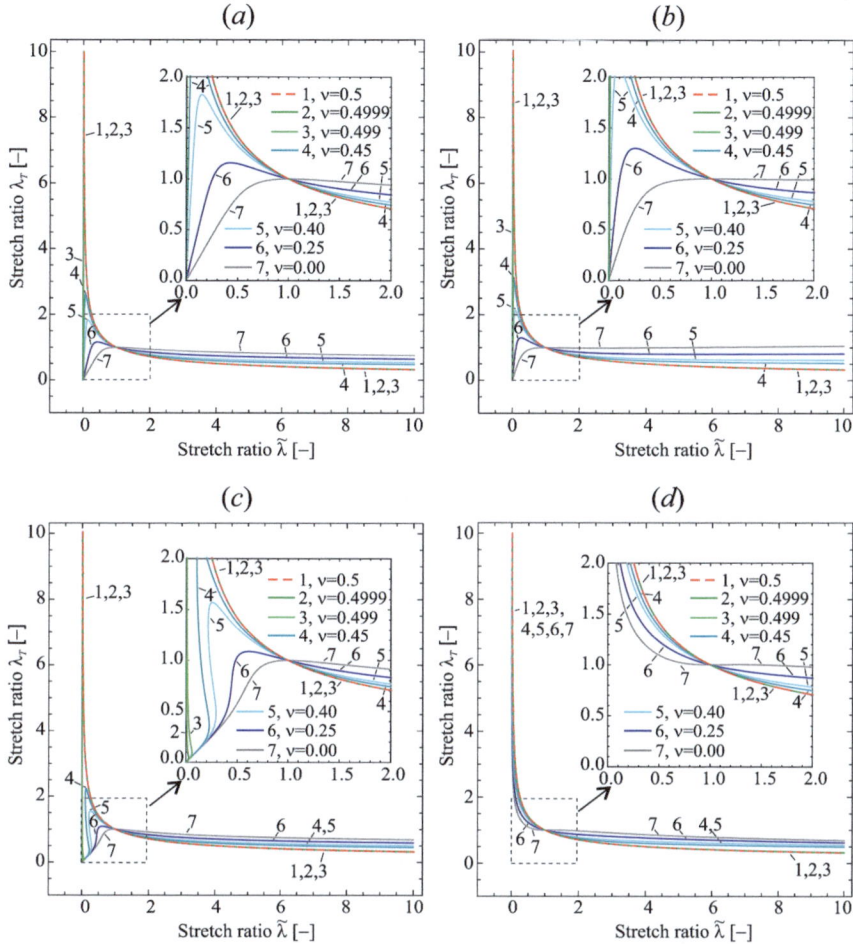

Fig. 6.9 Plots of λ_T versus $\tilde{\lambda}$ in the UL problem for vol-iso models #5 **a**, #6 **b**, #7 **c**, and #8 **d**

Determining the Lagrange multiplier p from $(6.9)_2$ and $(6.33)_2$ and using (6.32), from $(6.33)_1$ we obtain

$$\sigma_{11} = \mu\,(\tilde{\lambda}^2 - \tilde{\lambda}^{-4}). \tag{6.34}$$

Using (6.32) and (6.34), from (6.14) we get

$$P_{11} = \tilde{\lambda}^{-1}\sigma_{11} = \mu\,(\tilde{\lambda} - \tilde{\lambda}^{-5}). \tag{6.35}$$

Using expressions (6.32), (6.34), and (6.35), we obtain the limiting values of λ_T, σ_{11}, and P_{11} in extreme states, which coincide with the corresponding values for the UL problem (see Table 6.1).

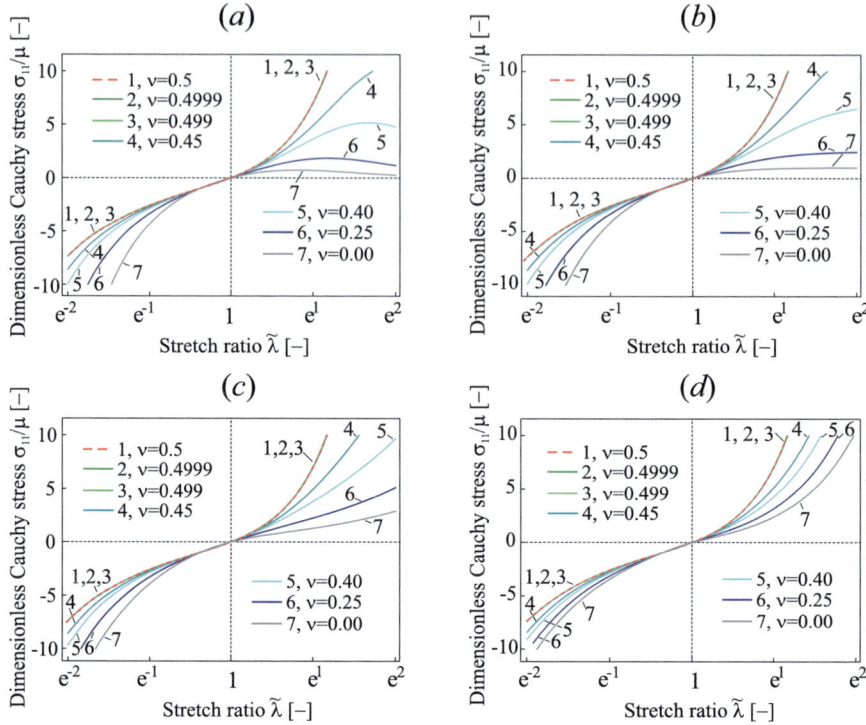

Fig. 6.10 Plots of σ_{11} versus $\tilde{\lambda}$ in the UL problem for vol-iso models #1 **a**, #2 **b**, #3 **c**, and #4 **d**

6.3.2 Compressible Isotropic Mixed Neo-Hookean Material Models

In view of (3.12) and (6.12), the components of the Cauchy stress tensor can be written as

$$\sigma_{11} = \sigma_{22} = \lambda\, h'(J) + \frac{\mu}{J}(\tilde{\lambda}^2 - 1), \qquad \sigma_{33} = \lambda\, h'(J) + \frac{\mu}{J}(\lambda_T^2 - 1). \tag{6.36}$$

Using $(6.9)_2$ and $(6.36)_2$, we obtain the implicit nonlinear dependence of λ_T on $\tilde{\lambda}$ in the general case:

$$\lambda\, h'(\tilde{\lambda}^2 \lambda_T) + \frac{\mu}{\tilde{\lambda}^2}(\lambda_T - \lambda_T^{-1}) = 0. \tag{6.37}$$

As in Sect. 6.2.2, we first consider the value $\nu = 0$ for Poisson's ratio. Since $\lambda = 0$ for this value of ν, from (6.37) we obtain equality (6.28), which does not depend on the choice of the volumetric function. Using (6.11), (6.14), and $(6.36)_1$, we get

$$\sigma_{11} = \sigma_{22} = \mu\,(1 - \tilde{\lambda}^{-2}), \qquad P_{11} = P_{22} = \mu\,(\tilde{\lambda} - \tilde{\lambda}^{-1}).$$

6.3 Equibiaxial Loading in Plane Stress

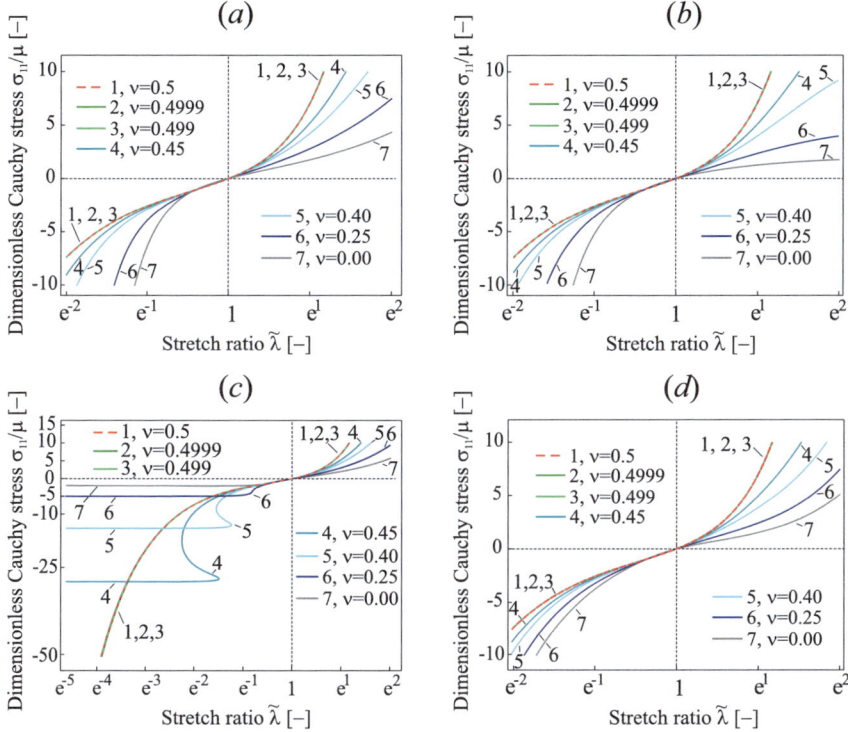

Fig. 6.11 Plots of σ_{11} versus $\tilde{\lambda}$ in the UL problem for vol-iso models #5 **a**, #6 **b**, #7 **c**, and #8 **d**

For the remaining values of ν from the interval $0 < \nu < 0.5$, the value of λ_T should be determined from the nonlinear equation (6.37). In the particular case of mixed model #7, using the function $h'(J)$ of the form $(4.5)_2$ in (6.37), we obtain the solution of Eq. (6.37) in closed form:

$$\lambda_T = \left[\mu/(\lambda \tilde{\lambda}^4 + \mu) \right]^{1/2}.$$

For the volumetric functions $h'(J)$ considered in this book, the dependence $\lambda_T(\tilde{\lambda})$ is derived from (6.37) using the Wolfram Mathematica software. Substitution of the obtained dependence into $(6.36)_1$ taking into account expression (6.11) yields the dependence $\sigma_{11}(\tilde{\lambda})$. The obtained dependencies $\lambda_T(\tilde{\lambda})$ and $\sigma_{11}(\tilde{\lambda})$ are used to derive expressions of $P_{11}(\tilde{\lambda}) = P_{22}(\tilde{\lambda})$ from (6.14).

Since the dependencies of out-of-plane principal stretch and stresses on prescribed in-plane principal stretches obtained in the ELP and ULP problems are qualitatively similar to the corresponding dependencies obtained in the UL problem and presented in Sect. 6.2, in this and subsequent sections, we restrict ourselves to testing models #1,4,7 for the most widely used volumetric functions and #8 for the new one. Plots

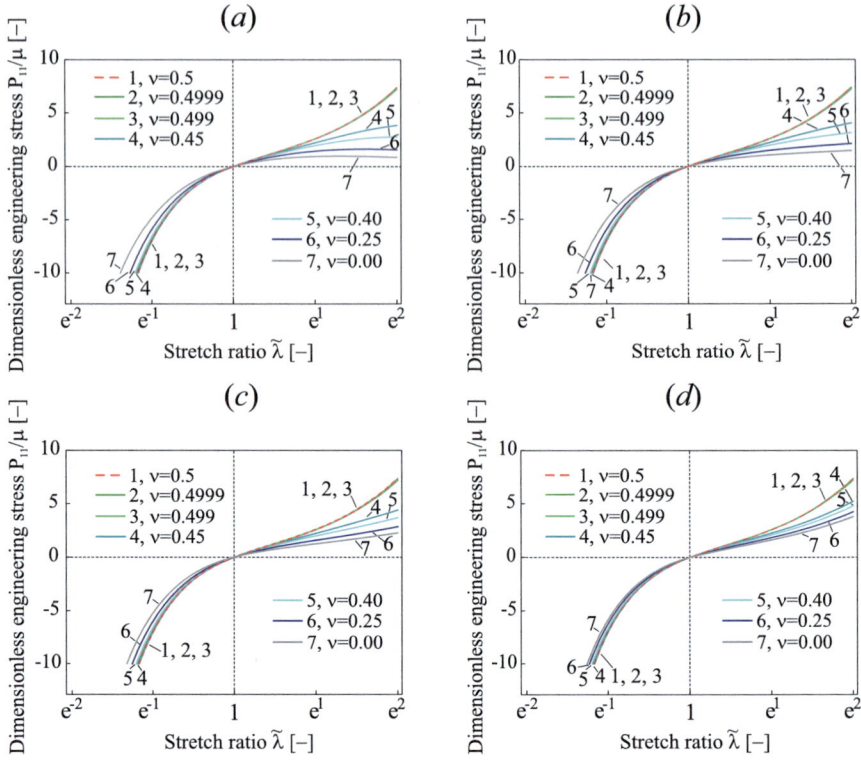

Fig. 6.12 Plots of P_{11} versus $\tilde{\lambda}$ in the UL problem for vol-iso models #1 **a**, #2 **b**, #3 **c**, and #4 **d**

of λ_T versus $\tilde{\lambda}$ are given in Fig. 6.14, and plots σ_{11} and P_{11} versus $\tilde{\lambda}$ are given in Figs. 6.15 and 6.16, respectively. We observe the non-monotonic physically inadmissible dependencies of Cauchy stresses on stretches in Fig. 6.16c for material model #7.

The limiting values of λ_T, σ_{11}, and P_{11} in extreme states are presented in Table 6.3. These limiting values coincide with the corresponding limiting values for the same material models in Table 6.2. Note that the limiting values of these quantities for model #7 coincide with the corresponding values for this material model in Table 2 in [4].

The conclusion following from the solutions of the ELP problem using mixed models is similar to the conclusion drawn from the analysis of solutions of the UL problem at the end of Sect. 6.2.2.

6.3 Equibiaxial Loading in Plane Stress

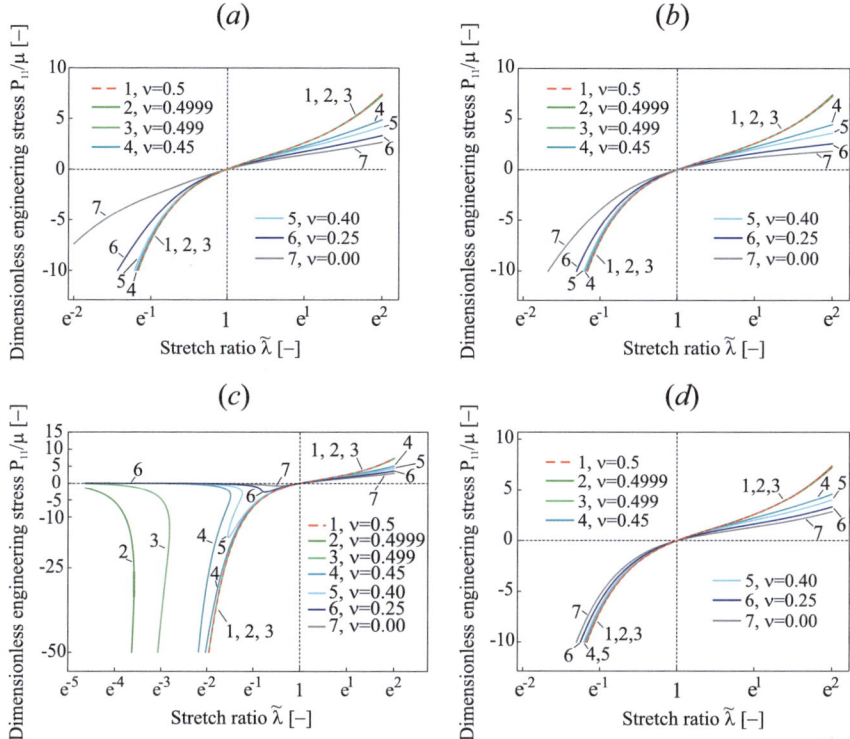

Fig. 6.13 Plots of P_{11} versus $\tilde{\lambda}$ in the UL problem for vol-iso models #5 **a**, #6 **b**, #7 **c**, and #8 **d**

Table 6.3 Limiting values of λ_T, σ_{11}, and P_{11} in extreme states where $\tilde{\lambda} \to 0$ and $\tilde{\lambda} \to \infty$ in the solution of the ELP problem for compressible isotropic material models with $0 \le \nu < 0.5$

Model ID	Quantity[b]	Mixed models		Vol-iso models	
		$\tilde{\lambda} \to 0$	$\tilde{\lambda} \to \infty$	$\tilde{\lambda} \to 0$	$\tilde{\lambda} \to \infty$
1	$\lambda_T(\tilde{\lambda})$	$+\infty$	0	**0**	$+\infty$
	$\sigma_{11}(\tilde{\lambda})$	$-\infty$	$+\infty$	$-\infty$	**0**
	$P_{11}(\tilde{\lambda})$	$-\infty$	$+\infty$	$-\infty$	**0**
4	$\lambda_T(\tilde{\lambda})$	$+\infty$	0	$+\infty$	0
	$\sigma_{11}(\tilde{\lambda})$	$-\infty$	$+\infty$	$-\infty$	$+\infty$
	$P_{11}(\tilde{\lambda})$	$-\infty$	$+\infty$	$-\infty$	$+\infty$
7	$\lambda_T(\tilde{\lambda})$	1	0	**0**	0
	$\sigma_{11}(\tilde{\lambda})$	$-\infty$	$+\infty$	$-3K/2$	$+\infty$
	$P_{11}(\tilde{\lambda})$	$-\infty$	$+\infty$	**0**	$+\infty$
8	$\lambda_T(\tilde{\lambda})$	$+\infty$	0	$+\infty$	0
	$\sigma_{11}(\tilde{\lambda})$	$-\infty$	$+\infty$	$-\infty$	$+\infty$
	$P_{11}(\tilde{\lambda})$	$-\infty$	$+\infty$	$-\infty$	$+\infty$

[b] Standard and highlighted texts indicate physically reasonable and unreasonable values of a quantity, respectively

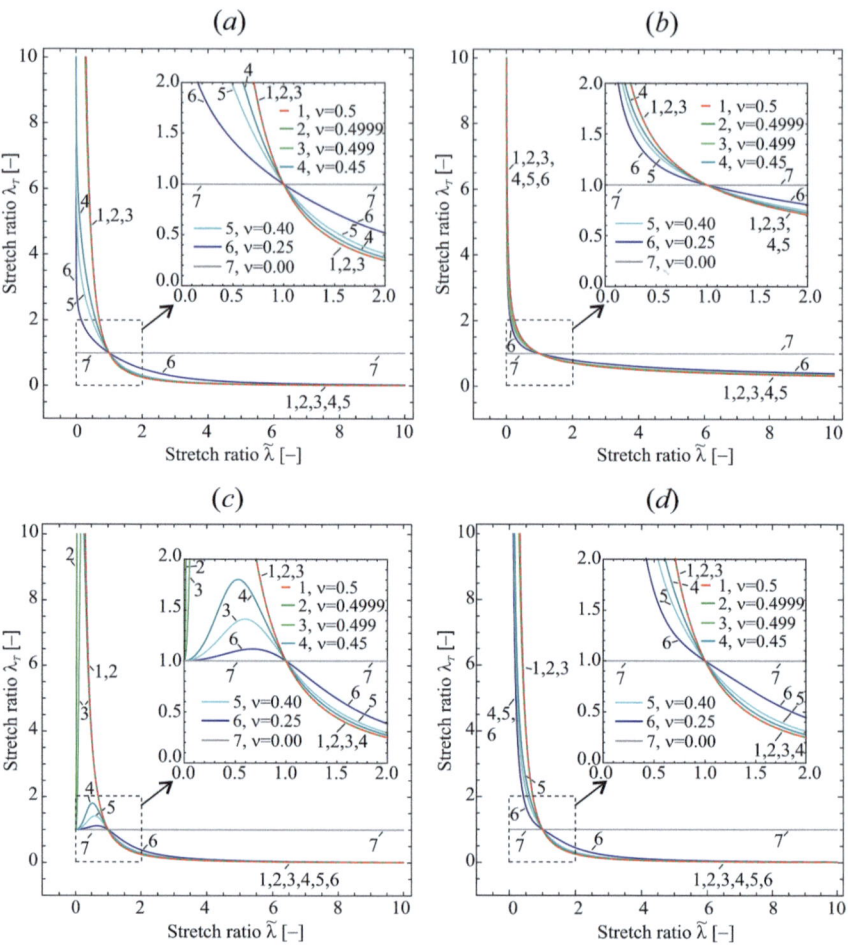

Fig. 6.14 Plots of λ_T versus $\tilde{\lambda}$ in the ELP problem for mixed models #1 **a**, #4 **b**, #7 **c**, and #8 **d**

6.3.3 Compressible Isotropic vol-iso Neo-Hookean Material Models

In view of (3.15) and (6.13), the components of the Cauchy stress tensor can be written as

$$\sigma_{11} = \sigma_{22} = K\, h'(J) + \frac{1}{3}\mu J^{-5/3}(\tilde{\lambda}^2 - \lambda_T^2), \qquad \sigma_{33} = K\, h'(J) + \frac{2}{3}\mu J^{-5/3}(\lambda_T^2 - \tilde{\lambda}^2). \quad (6.38)$$

Regardless of the choice of Poisson's ratio $\nu \in [0, 0.5)$, Eqs. $(6.9)_2$, (6.11), and $(6.38)_2$ lead to the following nonlinear equation for the dependence $\lambda_T(\tilde{\lambda})$:

6.3 Equibiaxial Loading in Plane Stress

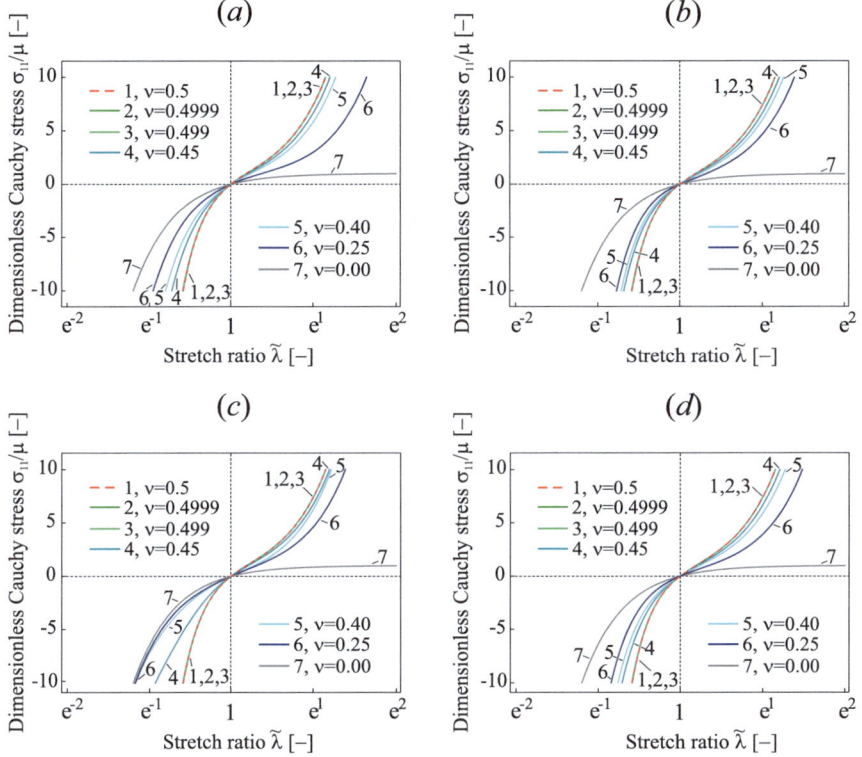

Fig. 6.15 Plots of σ_{11} versus $\tilde{\lambda}$ in the ELP problem for mixed models #1 **a**, #4 **b**, #7 **c**, and #8 **d**

$$K h'(\tilde{\lambda}^2 \lambda_T) + \frac{2}{3}\mu J^{-5/3}(\lambda_T^2 - \tilde{\lambda}^2) = 0.$$

Summing the left and right sides of the equalities in (6.38) and taking into account equality $\sigma_{33} = 0$, we get

$$\sigma_{11} = \sigma_{22} = \frac{3}{2} K h'(\tilde{\lambda}^2 \lambda_T). \tag{6.39}$$

Substitution of $\lambda_T(\tilde{\lambda})$ into the right-hand side of (6.39) leads to the dependencies $\sigma_{11}(\tilde{\lambda}) = \sigma_{22}(\tilde{\lambda})$. The dependencies $P_{11}(\tilde{\lambda}) = P_{22}(\tilde{\lambda})$ are obtained from (6.14) using the dependencies $\lambda_T(\tilde{\lambda})$ and $\sigma_{11}(\tilde{\lambda}) = \sigma_{22}(\tilde{\lambda})$.

Plots of λ_T versus $\tilde{\lambda}$ are shown in Fig. 6.17, and plots of σ_{11} and P_{11} versus $\tilde{\lambda}$ are given in Figs. 6.18 and 6.19, respectively. Note that the plots of $\lambda_T(\tilde{\lambda})$ in Fig. 6.17c, the plots of $\sigma_{11}(\tilde{\lambda})$ in Fig. 6.18c, and the plots of $P_{11}(\tilde{\lambda})$ in Fig. 6.19c for vol-iso material model #7 agree with the plots in Figs. 6, 14b, and 7a in [1] for the same material model.

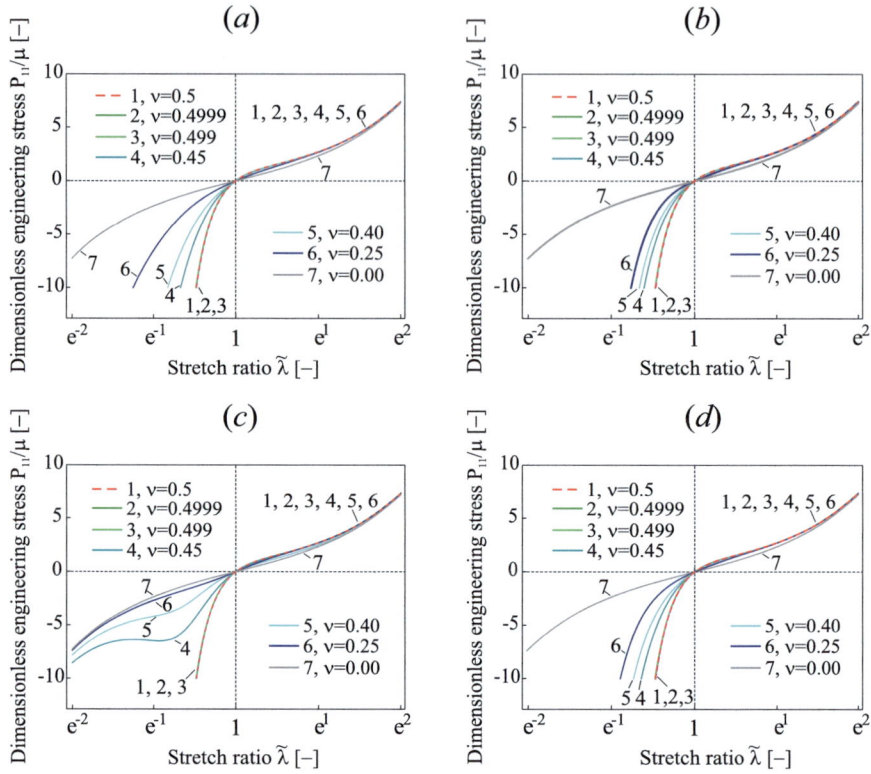

Fig. 6.16 Plots of P_{11} versus $\tilde{\lambda}$ in the ELP problem for mixed models #1 **a**, #4 **b**, #7 **c**, and #8 **d**

The limiting values of λ_T, σ_{11}, and P_{11} in extreme states are presented in Table 6.3. These limiting values are in qualitative agreement with the limiting values for the same material models presented in Table 6.2. Note that the limiting values of these quantities for model #3 are presented in Table 2 in [4]. The limiting values of these quantities for this model, though not presented in Table 6.3, agree with the limiting values in Table 2 in [4]. Note also that the limiting values of these quantities for model #7 coincide with those for the same model presented in Fig. 11 in [1].

The results of the solution of the ELP problem for vol-iso models lead to the conclusion that physically reasonable solutions can be obtained only using volumetric function #4 from the Hartmann–Neff family and new one #8.

6.4 Uniaxial Loading in Plane Strain

The dependencies of stresses and the unknown out-of-plane principal stretch on the prescribed longitudinal in-plane principal stretch obtained by solving the problem of uniaxial loading in plane strain for the incompressible isotropic neo-Hookean

6.4 Uniaxial Loading in Plane Strain

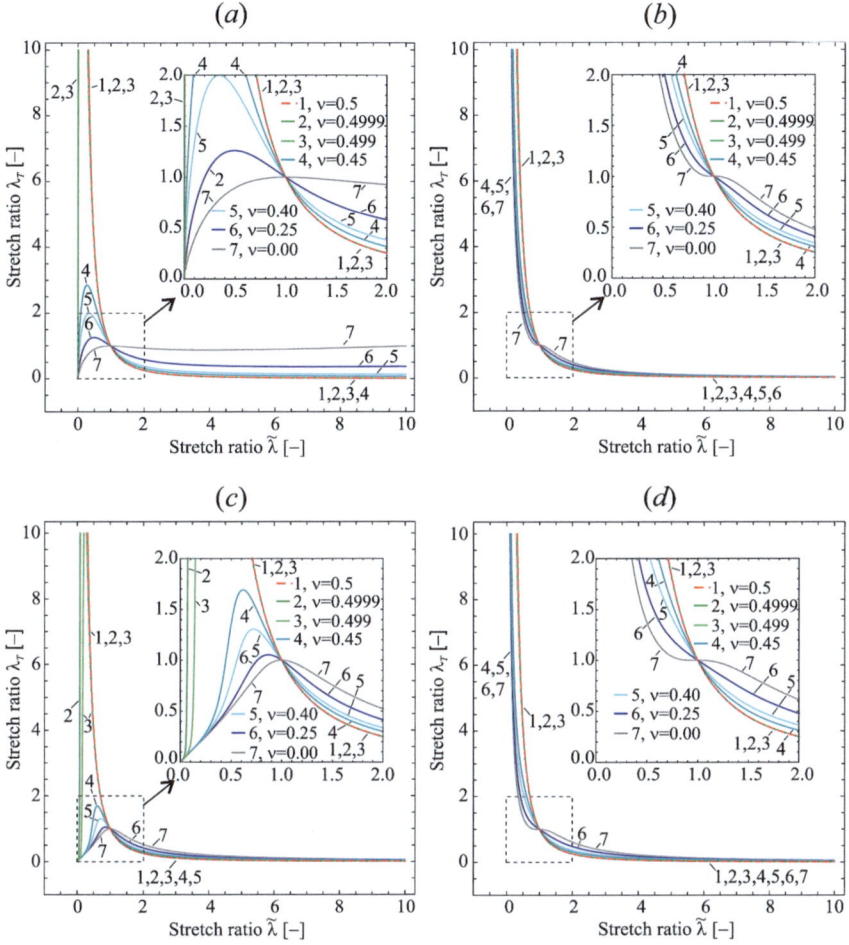

Fig. 6.17 Plots of λ_T versus $\tilde{\lambda}$ in the ELP problem for vol-iso models #1 **a**, #4 **b**, #7 **c**, and #8 **d**

material model and compressible mixed and vol-iso isotropic neo-Hookean models are presented in Sects. 6.4.1, 6.4.2, and 6.4.3, respectively.

6.4.1 Incompressible Isotropic Neo-Hookean Material

Setting $J = 1$, from (6.17) we obtain

$$\lambda_T = \tilde{\lambda}^{-1}. \tag{6.40}$$

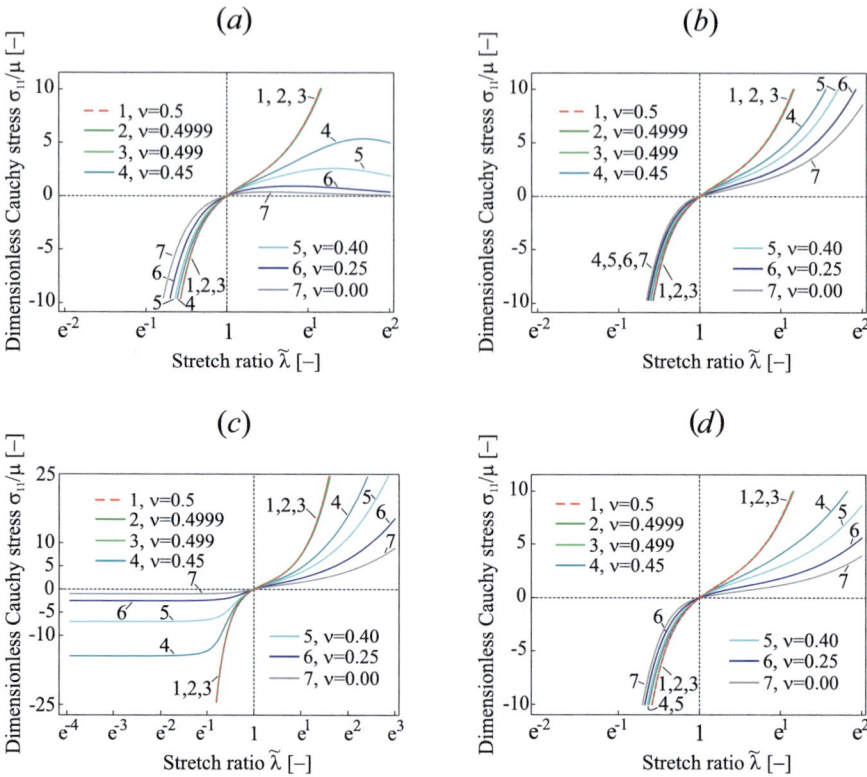

Fig. 6.18 Plots of σ_{11} versus $\tilde{\lambda}$ in the ELP problem for vol-iso models #1 **a**, #4 **b**, #7 **c**, and #8 **d**

In view of (3.4) and (6.18), the components of the Cauchy stress tensor can be written as

$$\sigma_{11} = \mu\,(\tilde{\lambda}^2 - 1) - p, \qquad \sigma_{22} = -p, \qquad \sigma_{33} = \mu\,(\lambda_T^2 - 1) - p. \qquad (6.41)$$

Determining the Lagrange multiplier p from $(6.15)_3$ and $(6.41)_3$ and using (6.40), from $(6.41)_{1,2}$ we obtain

$$\sigma_{11} = \mu\,(\tilde{\lambda}^2 - \tilde{\lambda}^{-2}), \qquad \sigma_{22} = -\mu\,(\tilde{\lambda}^{-2} - 1). \qquad (6.42)$$

Using (6.40) and (6.42), from (6.20) we get

$$P_{11} = \mu\,(\tilde{\lambda} - \tilde{\lambda}^{-3}), \qquad P_{22} = -\mu\,(\tilde{\lambda}^{-2} - 1). \qquad (6.43)$$

The limiting values of $\lambda_T, \sigma_{11}, \sigma_{22}, P_{11}$, and P_{22} in extreme states are obtained from expressions (6.40), (6.42), and (6.43) and are presented in Table 6.4. We assume that these limiting values correspond to the physically reasonable responses for idealized hyperelastic materials.

6.4 Uniaxial Loading in Plane Strain

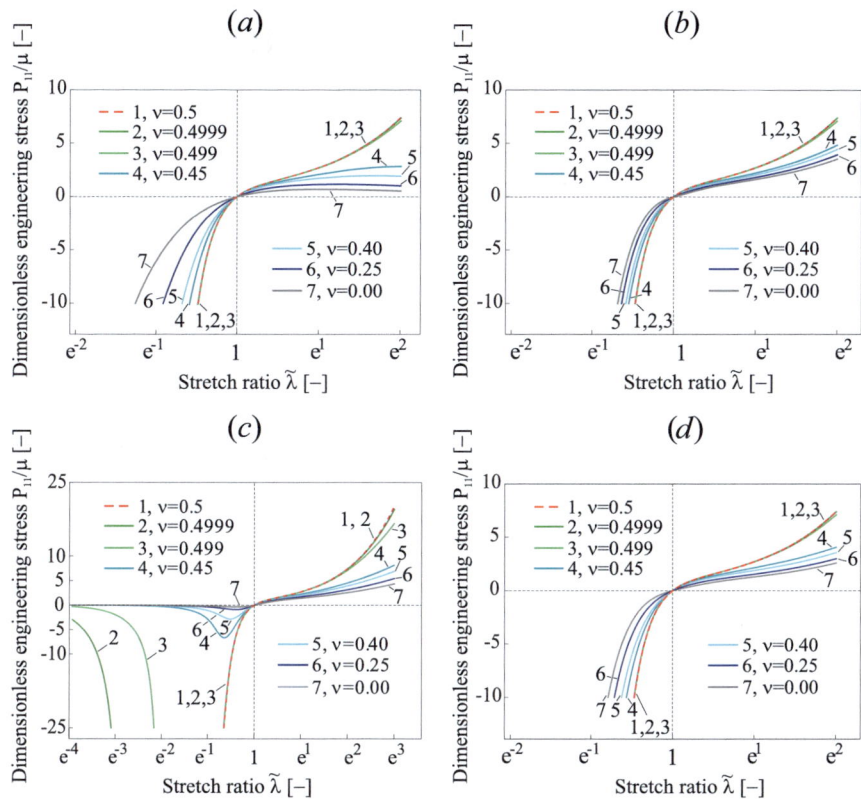

Fig. 6.19 Plots of P_{11} versus $\tilde{\lambda}$ in the ELP problem for vol-iso models #1 **a**, #4 **b**, #7 **c**, and #8 **d**

Table 6.4 Limiting values of λ_T, σ_{11}, σ_{22}, P_{11}, and P_{22} in extreme states where $\tilde{\lambda} \to 0$ and $\tilde{\lambda} \to \infty$ in the ULP problem for the incompressible isotropic neo-Hookean material model

Quantity	$\tilde{\lambda} \to 0$	$\tilde{\lambda} \to \infty$
$\lambda_T(\tilde{\lambda})$	$+\infty$	0
$\sigma_{11}(\tilde{\lambda})$	$-\infty$	$+\infty$
$\sigma_{22}(\tilde{\lambda})$	$-\infty$	$+\mu$
$P_{11}(\tilde{\lambda})$	$-\infty$	$+\infty$
$P_{22}(\tilde{\lambda})$	$-\infty$	$+\mu$

6.4.2 Compressible Isotropic Mixed Neo-Hookean Material Models

In view of (3.12) and (6.18), the components of the Cauchy stress tensor can be written as

$$\sigma_{11} = \lambda h'(J) + \frac{\mu}{J}(\tilde{\lambda}^2 - 1), \quad \sigma_{22} = \lambda h'(J), \quad \sigma_{33} = \lambda h'(J) + \frac{\mu}{J}(\lambda_T^2 - 1). \tag{6.44}$$

Using (6.15)$_3$ and (6.44)$_3$, we obtain the nonlinear implicit dependence of λ_T on $\tilde{\lambda}$ in the general case:

$$\lambda h'(\tilde{\lambda}\lambda_T) + \frac{\mu}{\tilde{\lambda}}(\lambda_T - \lambda_T^{-1}) = 0. \tag{6.45}$$

As in Sect. 6.2.2, we first consider the value $\nu = 0$ for Poisson's ratio. Since $\lambda = 0$ for this value of ν, from (6.45) we obtain equality (6.28), which does not depend on the choice of the volumetric function. Using (6.17), (6.20), and (6.44)$_{1,2}$, we get

$$\sigma_{11} = \mu(\tilde{\lambda} - \tilde{\lambda}^{-1}), \quad \sigma_{22} = 0, \quad P_{11} = \sigma_{11} = \mu(\tilde{\lambda} - \tilde{\lambda}^{-1}), \quad P_{22} = 0.$$

For the remaining values of ν from the interval $0 < \nu < 0.5$, the value of λ_T should be determined from the nonlinear equation (6.45). In the particular case of mixed model #7, using the function $h'(J)$ of the form (4.5)$_2$ in (6.45), we obtain the solution of Eq. (6.45) in closed form:

$$\lambda_T = \frac{1}{2(\lambda\tilde{\lambda}^2 + \mu)}(\lambda\tilde{\lambda} + \sqrt{\lambda^2\tilde{\lambda}^2 + 4\mu(\lambda\tilde{\lambda}^2 + \mu)}).$$

For all remaining volumetric functions $h'(J)$ considered in this book, the dependence $\lambda_T(\tilde{\lambda})$ is derived from (6.45) using the Wolfram Mathematica software. Substitution of the obtained dependence into (6.44)$_{1,2}$ taking into account expression (6.17) yields the dependencies $\sigma_{11}(\tilde{\lambda})$ and $\sigma_{22}(\tilde{\lambda})$. The obtained dependencies $\lambda_T(\tilde{\lambda})$, $\sigma_{11}(\tilde{\lambda})$, and $\sigma_{22}(\tilde{\lambda})$ are used to derive expressions for $P_{11}(\tilde{\lambda})$ and $P_{22}(\tilde{\lambda})$ from (6.20).

Plots of λ_T versus $\tilde{\lambda}$ are given in Fig. 6.20 and plots of σ_{11}, σ_{22}, P_{11}, and P_{22} versus $\tilde{\lambda}$ are shown in Figs. 6.21, 6.22, 6.23 and 6.24, respectively. The limiting values of λ_T, σ_{11}, σ_{22}, P_{11}, and P_{22} in extreme states are presented in Table 6.5. These limiting values mostly coincide with the limiting values for the same material models presented in Table 6.4.

The conclusion following from the solutions of the ULP problem for mixed models is similar to the conclusion drawn from the analysis of solutions of the UL problem at the end of Sect. 6.2.2.

6.4 Uniaxial Loading in Plane Strain

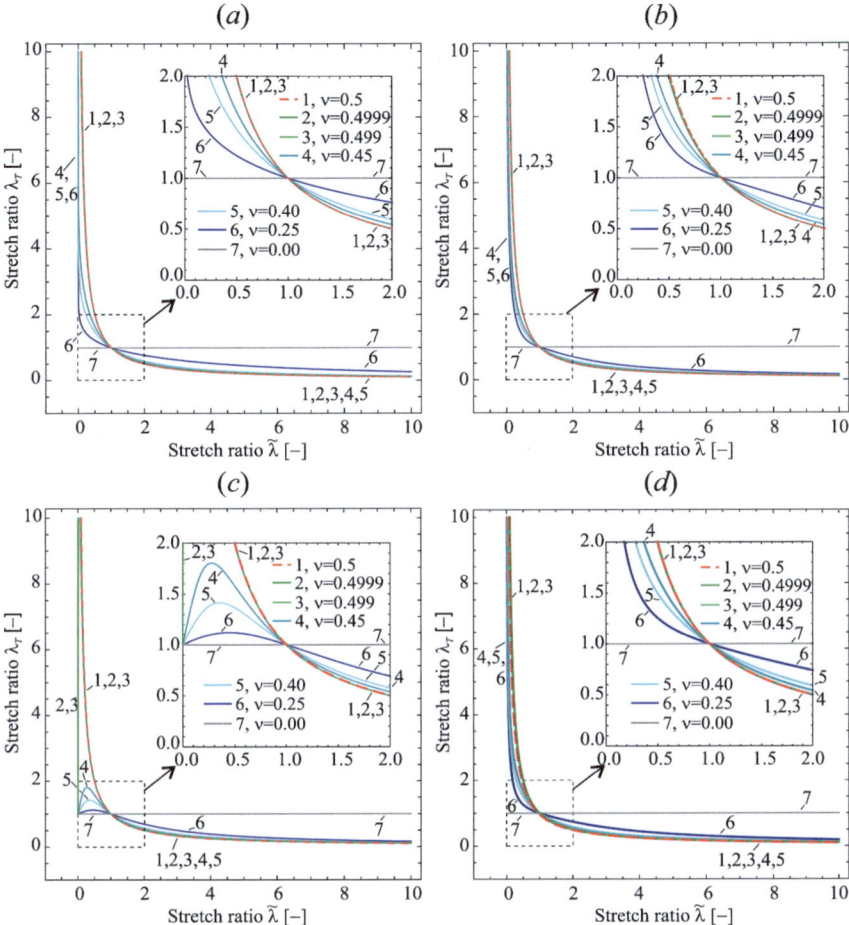

Fig. 6.20 Plots of λ_T versus $\tilde{\lambda}$ in the ULP problem for mixed models #1 **a**, #4 **b**, #7 **c**, and #8 **d**

6.4.3 Compressible Isotropic vol-iso Neo-Hookean Material Models

In view of (3.15) and (6.19), the components of the Cauchy stress tensor can be written as

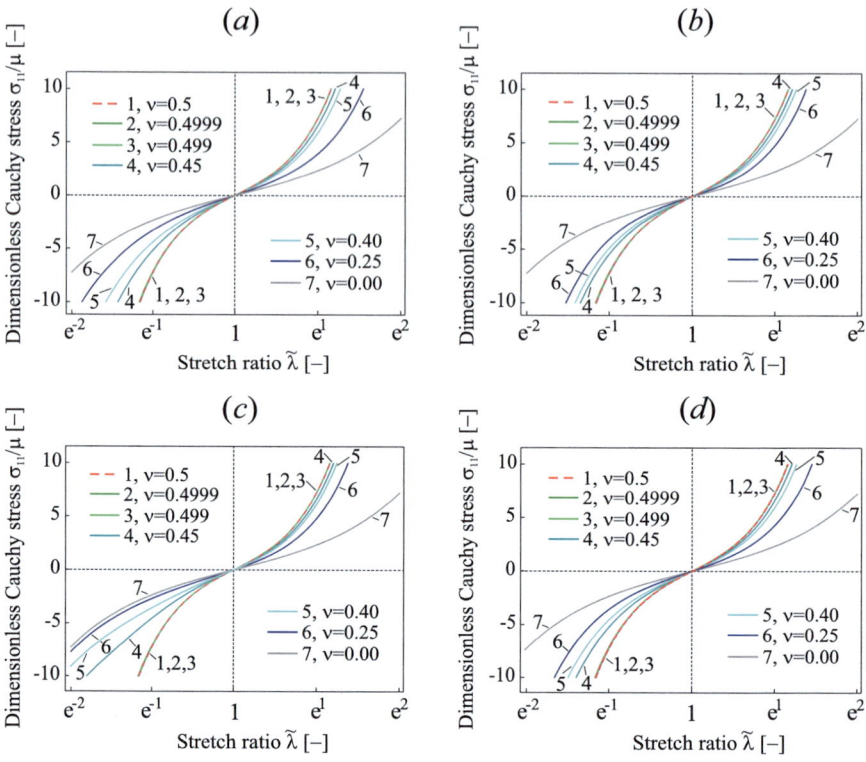

Fig. 6.21 Plots of σ_{11} versus $\tilde{\lambda}$ in the ULP problem for mixed models #1 **a**, #4 **b**, #7 **c**, and #8 **d**

$$\sigma_{11} = K h'(J) + \frac{1}{3}\mu J^{-5/3}(2\tilde{\lambda}^2 - 1 - \lambda_T^2), \tag{6.46}$$

$$\sigma_{22} = K h'(J) + \frac{1}{3}\mu J^{-5/3}(-\tilde{\lambda}^2 + 2 - \lambda_T^2),$$

$$\sigma_{33} = K h'(J) + \frac{1}{3}\mu J^{-5/3}(2\lambda_T^2 - 1 - \tilde{\lambda}^2).$$

Regardless of the choice of Poisson's ratio $\nu \in [0, 0.5)$, Eqs. (6.15)$_3$, (6.17), and (6.46)$_3$ lead to the following nonlinear equation for the dependence $\lambda_T(\tilde{\lambda})$:

$$K h'(\tilde{\lambda}\lambda_T) + \frac{1}{3}\mu J^{-5/3}(2\lambda_T^2 - 1 - \tilde{\lambda}^2) = 0.$$

Substitution of the dependencies $\lambda_T(\tilde{\lambda})$ into the right-hand side of (6.46)$_{1,2}$ yields the dependencies $\sigma_{11}(\tilde{\lambda})$ and $\sigma_{22}(\tilde{\lambda})$. The dependencies $P_{11}(\tilde{\lambda})$ and $P_{22}(\tilde{\lambda})$ are obtained from (6.20) using the dependencies $\lambda_T(\tilde{\lambda})$, $\sigma_{11}(\tilde{\lambda})$, and $\sigma_{22}(\tilde{\lambda})$.

6.4 Uniaxial Loading in Plane Strain

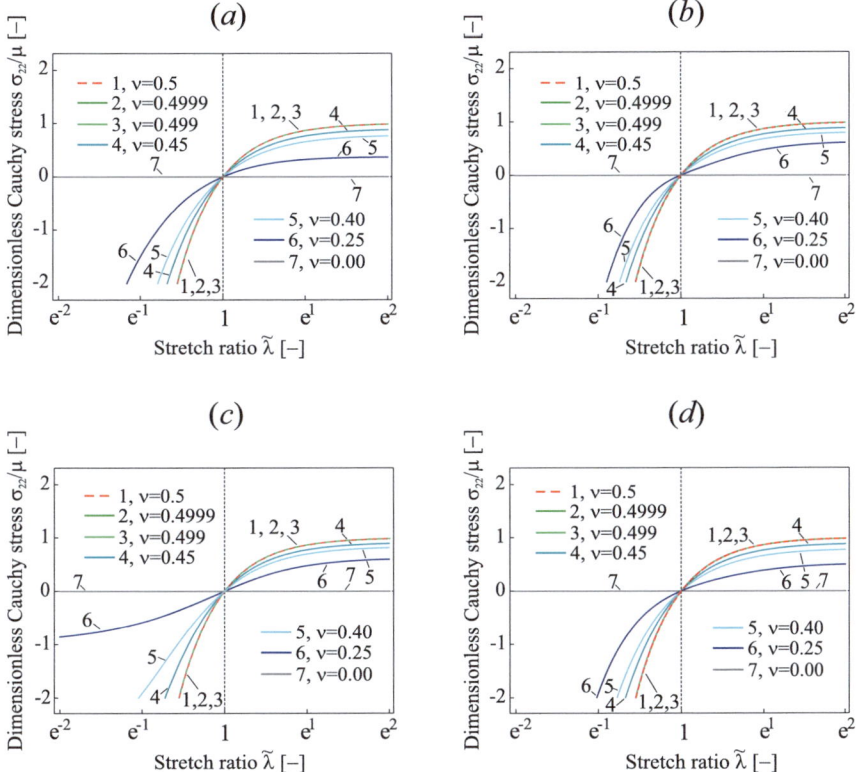

Fig. 6.22 Plots of σ_{22} versus $\tilde{\lambda}$ in the ULP problem for mixed models #1 **a**, #4 **b**, #7 **c**, and #8 **d**

Plots of λ_T versus $\tilde{\lambda}$ are shown in Fig. 6.25, and plots of σ_{11}, σ_{22}, P_{11}, and P_{22} versus $\tilde{\lambda}$ in Figs. 6.26, 6.27, 6.28 and 6.29, respectively. Note that the plots of $\lambda_T(\tilde{\lambda})$ in Fig. 6.25c, the plots of $\sigma_{11}(\tilde{\lambda})$ in Fig. 6.26c, and the plots of $P_{11}(\tilde{\lambda})$ in Fig. 6.28c for vol-iso material model #7 agree with the plots in Figs. 8, 14c, and 9a in [1] for the same material model. We observe the non-monotonic physically inadmissible dependencies of Cauchy stresses on stretches in Fig. 6.27c for material model #7.

The limiting values of λ_T, σ_{11}, σ_{22}, P_{11}, and P_{22} in extreme states are presented in Table 6.5. These limiting values are in qualitative agreement with the limiting values for the same material models presented in Table 6.3. Note that the limiting values of these quantities for model #7 coincide with those for the same model presented in Fig. 11 in [1].

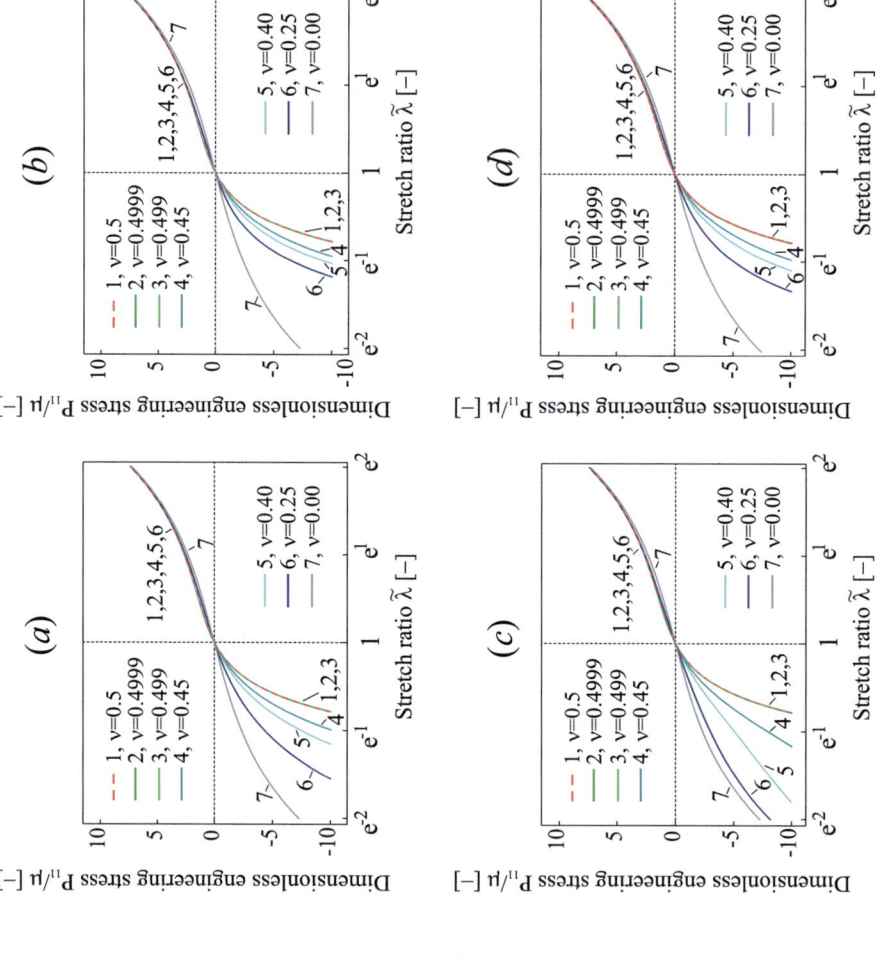

Fig. 6.23 Plots of P_{11} versus $\tilde{\lambda}$ in the ULP problem for mixed models #1 **a**, #4 **b**, #7 **c**, and #8 **d**

6.4 Uniaxial Loading in Plane Strain

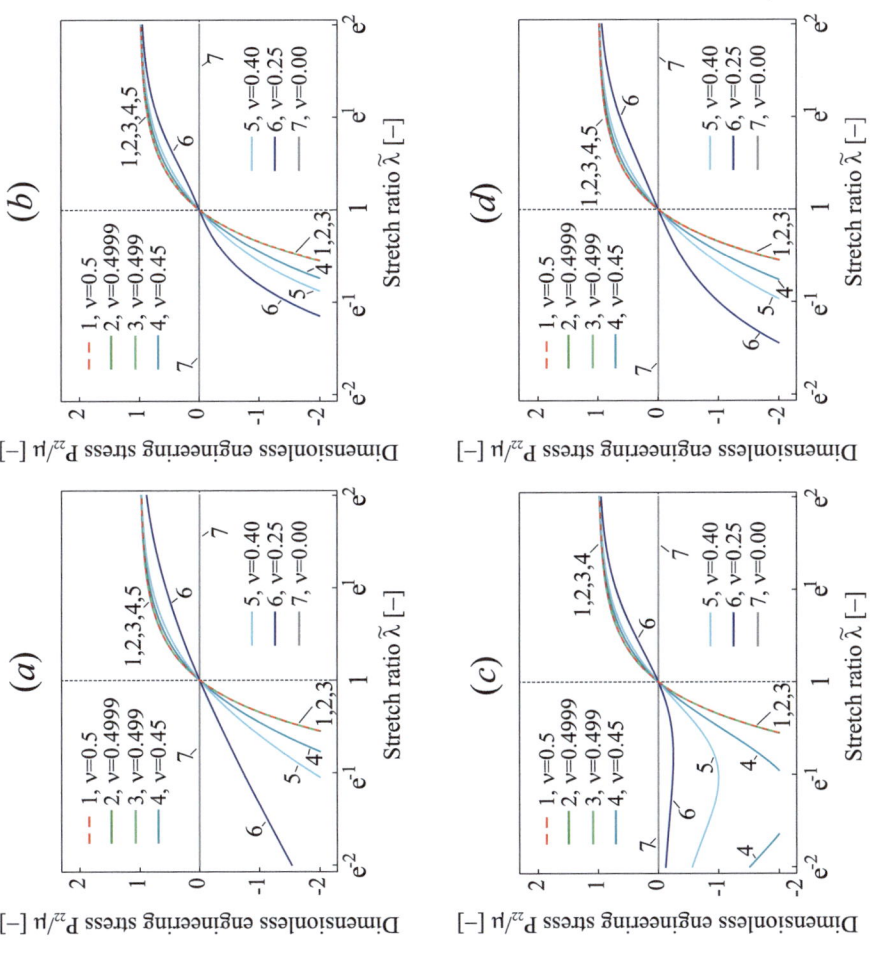

Fig. 6.24 Plots of P_{22} versus $\tilde{\lambda}$ in the ULP problem for mixed models #1 **a**, #4 **b**, #7 **c**, and #8 **d**

Table 6.5 Limiting values of $\lambda_T, \sigma_{11}, \sigma_{22}, P_{11}$, and P_{22} in extreme states where $\tilde{\lambda} \to 0$ and $\tilde{\lambda} \to \infty$ in the solution of the ULP problem for the compressible isotropic material models with $0 \le \nu < 0.5$

Model ID	Quantity[b]	Mixed models		Vol-iso models	
		$\tilde{\lambda} \to 0$	$\tilde{\lambda} \to \infty$	$\tilde{\lambda} \to 0$	$\tilde{\lambda} \to \infty$
1	$\lambda_T(\tilde{\lambda})$	$+\infty$	0	$1/\sqrt{2}$	$+\infty$
	$\sigma_{11}(\tilde{\lambda})$	$-\infty$	$+\infty$	$-\infty$	**0**
	$\sigma_{22}(\tilde{\lambda})$	$-\infty$	$*$	$+\infty$	**0**
	$P_{11}(\tilde{\lambda})$	$-\infty$	$+\infty$	$-\infty$	**0**
	$P_{22}(\tilde{\lambda})$	$-\infty$	$+\mu$	$\pm\infty$	$-\infty$
4	$\lambda_T(\tilde{\lambda})$	$+\infty$	0	$+\infty$	0
	$\sigma_{11}(\tilde{\lambda})$	$-\infty$	$+\infty$	$-\infty$	$+\infty$
	$\sigma_{22}(\tilde{\lambda})$	$-\infty$	$*$	$-\infty$	$+\infty$
	$P_{11}(\tilde{\lambda})$	$-\infty$	$+\infty$	$-\infty$	$+\infty$
	$P_{22}(\tilde{\lambda})$	$-\infty$	$+\mu$	$-\infty$	$-\infty$
7	$\lambda_T(\tilde{\lambda})$	**1**	0	$1/\sqrt{2}$	0
	$\sigma_{11}(\tilde{\lambda})$	$-\infty$	$+\infty$	$-\infty$	$+\infty$
	$\sigma_{22}(\tilde{\lambda})$	$-\lambda$	$*$	$+\infty$	$+\infty$
	$P_{11}(\tilde{\lambda})$	$-\infty$	$+\infty$	$-\infty$	$+\infty$
	$P_{22}(\tilde{\lambda})$	**0**	$+\mu$	$+\infty$	$+\infty$
8	$\lambda_T(\tilde{\lambda})$	$+\infty$	0	$+\infty$	0
	$\sigma_{11}(\tilde{\lambda})$	$-\infty$	$+\infty$	$-\infty$	$+\infty$
	$\sigma_{22}(\tilde{\lambda})$	$-\infty$	$*$	$-\infty$	$\pm\infty$
	$P_{11}(\tilde{\lambda})$	$-\infty$	$+\infty$	$-\infty$	$+\infty$
	$P_{22}(\tilde{\lambda})$	$-\infty$	$+\mu$	$-\infty$	$\pm\infty$

[b] An asterisk ($*$) denotes some finite limiting values, standard and highlighted texts indicate physically reasonable and unreasonable values of a quantity, and a symbol $\pm\infty$ denotes limiting values $+\infty$ or $-\infty$ for different values of Poisson's ratio

The conclusion following from the solutions of the ULP problem using vol-iso models is similar to the conclusion drawn from the analysis of solutions of the ELP problem at the end of Sect. 6.3.3.

6.4 Uniaxial Loading in Plane Strain

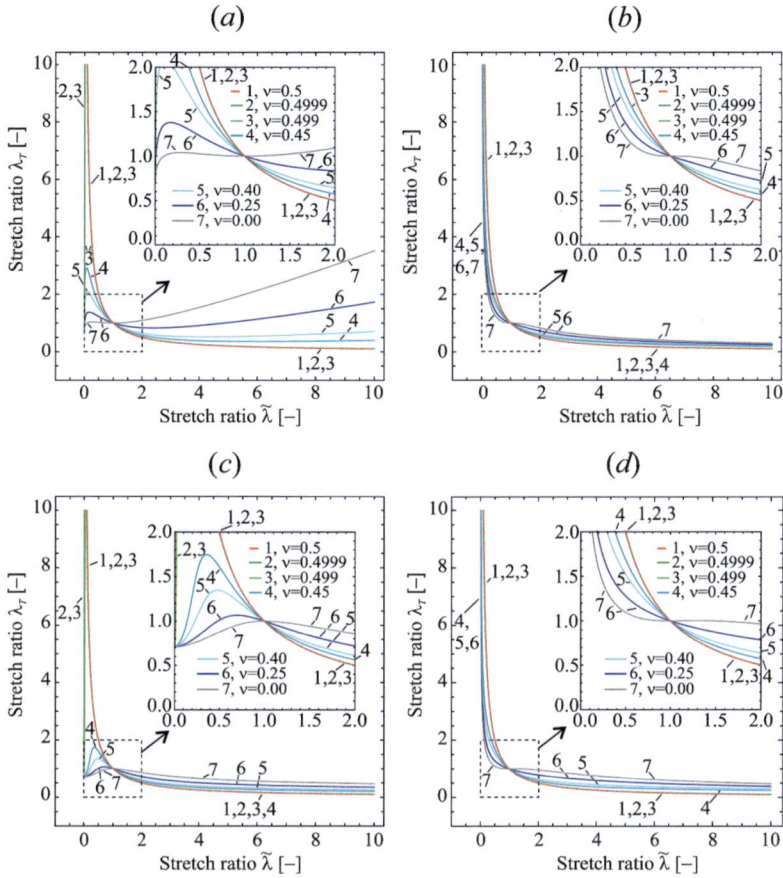

Fig. 6.25 Plots of λ_T versus $\tilde{\lambda}$ in the ULP problem for vol-iso models #1 **a**, #4 **b**, #7 **c**, and #8 **d**

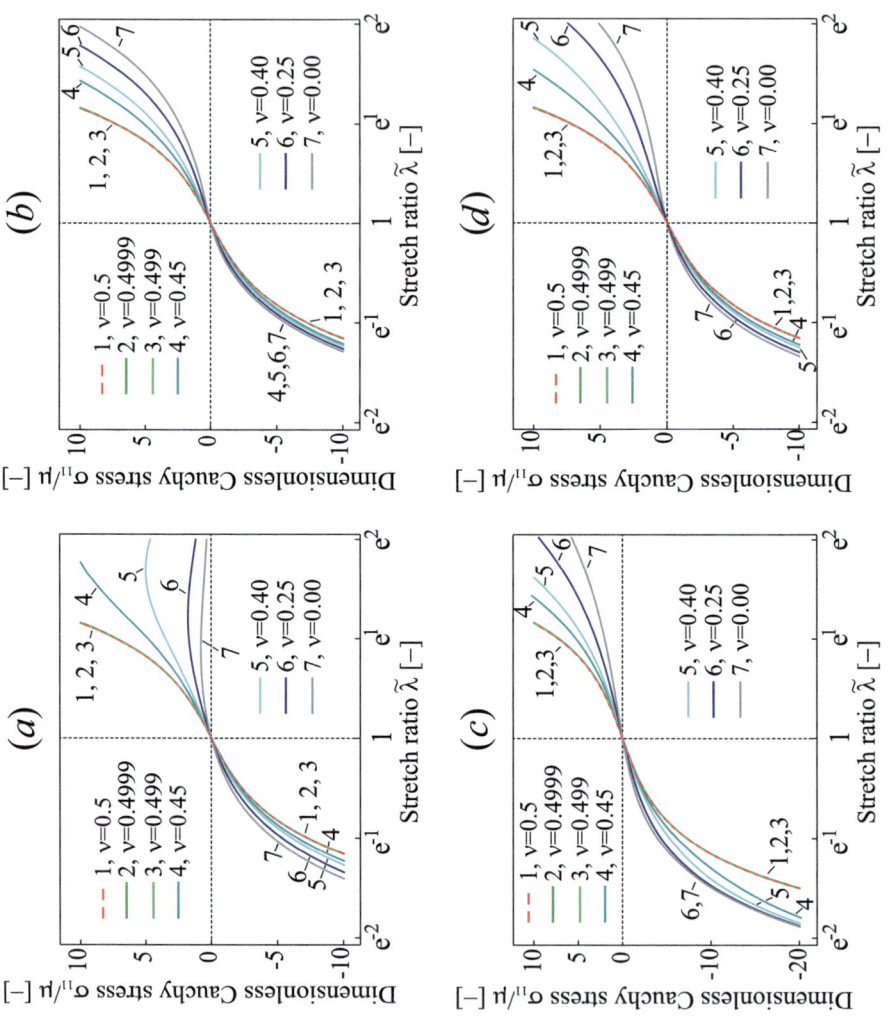

Fig. 6.26 Plots of σ_{11} versus $\tilde{\lambda}$ in the ULP problem for vol-iso models #1 **a**, #4 **b**, #7 **c**, and #8 **d**

6.4 Uniaxial Loading in Plane Strain

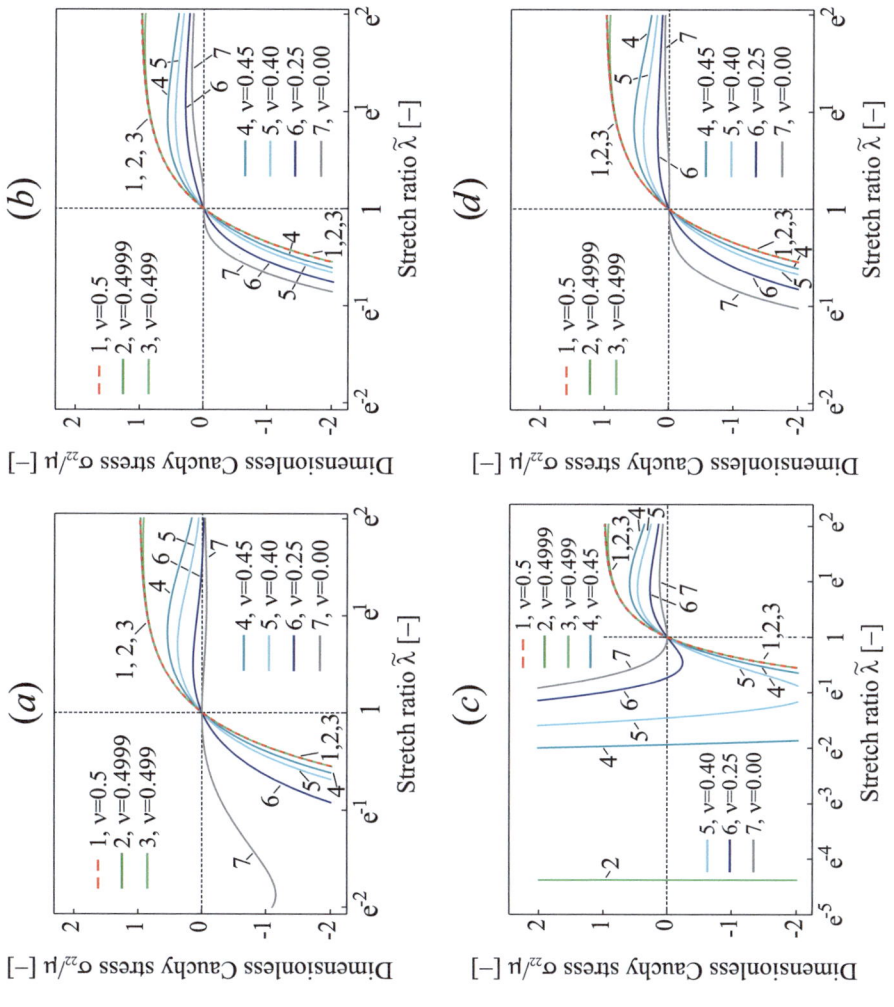

Fig. 6.27 Plots of σ_{22} versus $\tilde{\lambda}$ in the ULP problem for vol-iso models #1 **a**, #4 **b**, #7 **c**, and #8 **d**

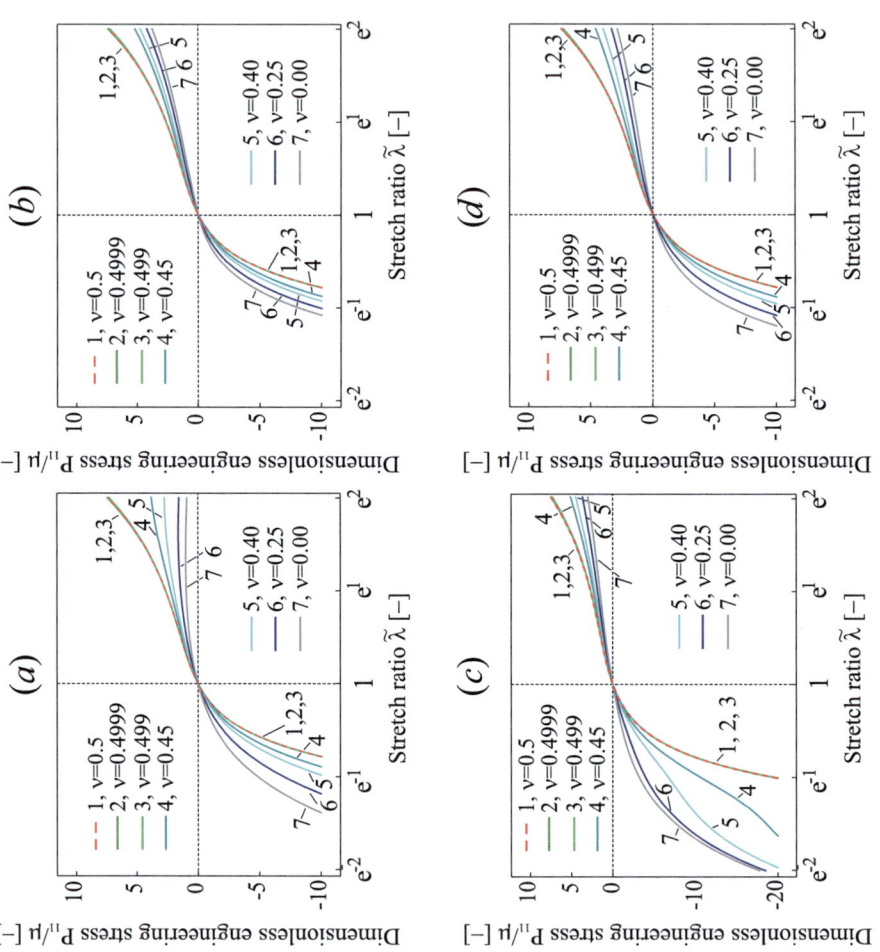

Fig. 6.28 Plots of P_{11} versus $\tilde{\lambda}$ in the ULP problem for vol-iso models #1 **a**, #4 **b**, #7 **c**, and #8 **d**

6.4 Uniaxial Loading in Plane Strain

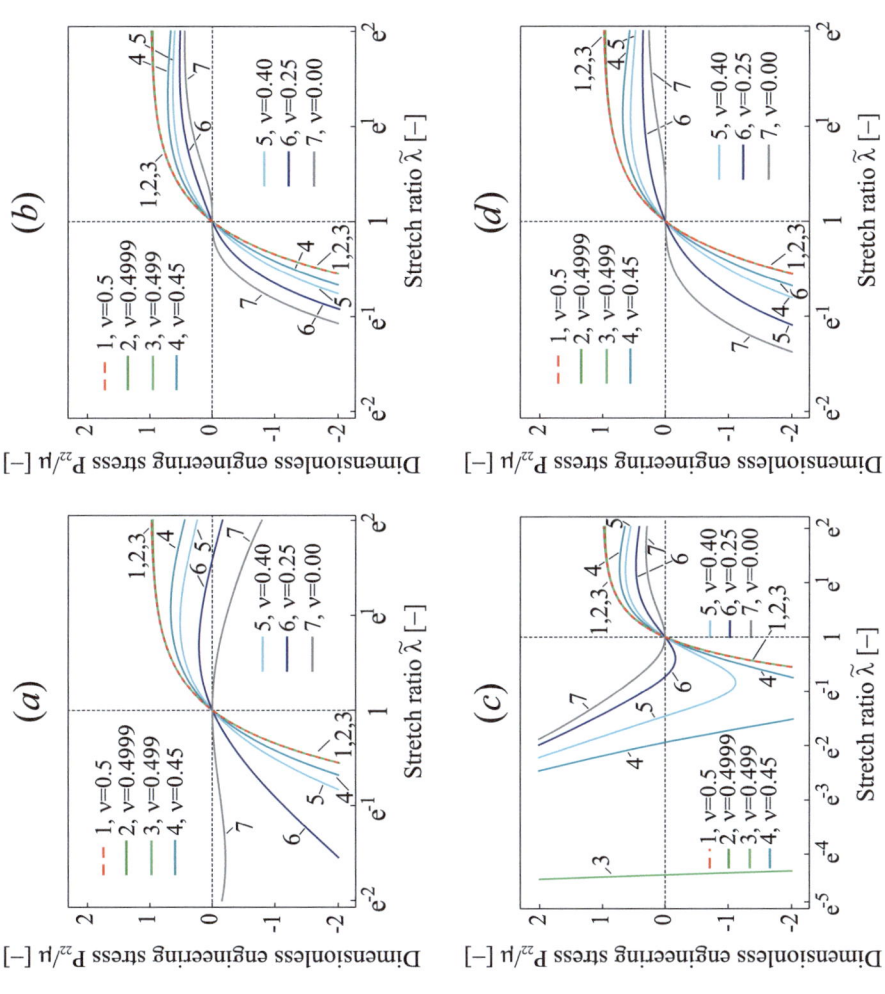

Fig. 6.29 Plots of P_{22} versus $\tilde{\lambda}$ in the ULP problem for vol-iso models #1 **a**, #4 **b**, #7 **c**, and #8 **d**

References

1. Kossa, A., Valentine, M.T., McMeeking, R.M.: Analysis of the compressible, isotropic, neo-Hookean hyperelastic model. Meccanica **58**(1), 217–232 (2023). https://doi.org/10.1007/s11012-022-01633-2
2. Hartmann, S., Neff, P.: Polyconvexity of generalized polynomial-type hyperelastic strain energy functions for near-incompressibility. Int. J. Solids Struct. **40**(11), 2767–2791 (2003). https://doi.org/10.1016/S0020-7683(03)00086-6
3. Kellermann, D.C., Attard, M.M.: An invariant-free formulation of neo-Hookean hyperelasticity. Zeitschrift für Angewandte Mathematik und Mechanik **96**(2), 233–252 (2016). https://doi.org/10.1002/zamm.201400210
4. Pence, T.J., Gou, K.: On compressible versions of the incompressible neo-Hookean material. Math. Mech. Solids **20**(2), 157–182 (2015). https://doi.org/10.1177/1081286514544258
5. Ehlers, W., Eipper, G.: The simple tension problem at large volumetric strains computed from finite hyperelastic material laws. Acta Mechanica **130**(1), 17–27 (1998). https://doi.org/10.1007/BF01187040

Chapter 7
Concluding Remarks

Abstract In this chapter we summarize and comment on all the research presented in this book. In particular, we do a comparative analysis of the performance of mixed and vol-iso material models in Sect. 7.1 and general conclusions in Sect. 7.2.

7.1 Comparative Analysis of the Performance of Mixed and vol-iso Material Models

The analysis performed in this book leads to the following conclusions.

1. From the application point of view, both mixed and vol-iso models predict close values of stress tensors and principal stretches for slightly compressible materials. However, for the models to be able to predict the dilatation $J - 1$, they should be somewhat modified (see, e.g., [1, 2]), which is beyond the scope of the present study.
2. From the point of view of the consistency of physical intuition, in limiting states (i.e., for extreme values of stretches, very large or very small), models #3,4,8 of both types (mixed and vol-iso) have physically reasonable responses for kinematic and static quantities. For models #3,4 we use volumetric functions from the Hartmann–Neff family with parameter $q = 2, 5$ and for model #8 we use new volumetric function. However, for model #1 with the volumetric function with $q = 0$ (i.e., with the function $h^{(0)} = (\ln J)^2/2$), only the mixed model has physically reasonable responses.
3. Both mixed and vol-iso models #1–6,8 satisfy the Hill inequality, but model #7 do not satisfy this inequality for some values of the volume ratio J. In addition, all mixed and vol-iso models do not satisfy the CSP.
4. The constitutive relations for mixed models are more simpler than the constitutive relations for vol-iso models. This is especially true for rate formulations for these models (cf., expression (5.13) for mixed models and expression (5.19) for vol-iso ones). In this book, we used rate formulations of constitutive relations to answer the question of whether the material models considered here satisfy the Hill and corotational stability postulates. However, rate formulations of constitutive

relations are also required to implement any material models in FE systems. Our next goal is to obtain explicit expressions for tangent stiffness tensors for the material models under consideration.

The rate formulations of constitutive relations (5.13) for mixed models can be rewritten as

$$\frac{D^{ZJ}}{Dt}[\tau] = \lambda \chi(J) J \operatorname{tr} \mathbf{d} \mathbf{I} + 2(\mu - \lambda \ln J)\mathbf{d} + \mathbf{d} \cdot \tau + \tau \cdot \mathbf{d}. \tag{7.1}$$

Similarly, the rate formulations of constitutive relations for vol-iso models (5.19) can be rewritten as

$$\frac{D^{ZJ}}{Dt}[\tau] = [K \chi(J)J + \frac{2}{9}\mu \operatorname{tr} \mathbf{c} \, J^{-2/3}] \operatorname{tr} \mathbf{d} \mathbf{I} + [\frac{2}{3}\mu \operatorname{tr} \mathbf{c} \, J^{-2/3} - 2K \, Jh'(J)]\mathbf{d}$$
$$- \frac{2}{3}\mu \, J^{-2/3}[\operatorname{tr} \mathbf{d} \mathbf{c} + (\mathbf{c} : \mathbf{d})\mathbf{I}] + \mathbf{d} \cdot \tau + \tau \cdot \mathbf{d}. \tag{7.2}$$

Alternative forms of the rate constitutive relations (7.1) and (7.2) can be obtained using the upper Oldroyd stress rate $\frac{D^{\overline{Old}}}{Dt}[\tau]$ instead of the Zaremba–Jaumann stress rate $\frac{D^{ZJ}}{Dt}[\tau]$. Based on the relationship between objective tensor rates (see, e.g., Eq. $(25)_2$ in [3] and Eq. (47) in [4]))

$$\frac{D^{\overline{Old}}}{Dt}[\tau] = \frac{D^{ZJ}}{Dt}[\tau] - \mathbf{d} \cdot \tau - \tau \cdot \mathbf{d}, \tag{7.3}$$

an alternative representation of the rate constitutive relations for mixed models can be obtained from (7.1):

$$\frac{D^{\overline{Old}}}{Dt}[\tau] = \lambda \chi(J) J \operatorname{tr} \mathbf{d} \mathbf{I} + 2(\mu - \lambda \ln J)\mathbf{d}, \tag{7.4}$$

and that for vol-iso models can be obtained from (7.2):

$$\frac{D^{\overline{Old}}}{Dt}[\tau] = [K \chi(J)J + \frac{2}{9}\mu \operatorname{tr} \mathbf{c} \, J^{-2/3}] \operatorname{tr} \mathbf{d} \mathbf{I} + [\frac{2}{3}\mu \operatorname{tr} \mathbf{c} \, J^{-2/3} - 2K \, Jh'(J)]\mathbf{d}$$
$$- \frac{2}{3}\mu \, J^{-2/3}[\operatorname{tr} \mathbf{d} \mathbf{c} + (\mathbf{c} : \mathbf{d})\mathbf{I}]. \tag{7.5}$$

In particular, for mixed model #1 (the *Simo–Pister* [5] *hyperelastic model*), the equality $\chi(J)J = 1$ holds and expression (7.4) reduces to the expression

$$\frac{D^{\overline{Old}}}{Dt}[\tau] = 2(\mu - \lambda \ln J)\mathbf{d} + \lambda \operatorname{tr} \mathbf{d} \mathbf{I},$$

7.1 Comparative Analysis of the Performance of Mixed ...

which represents the rate constitutive relations for the one-parameter (with parameter $n = 2$) family of Hooke-like isotropic hyper-/hypo-elastic material models (see Eq. (60) in [6] with the identification $\boldsymbol{\tau}^{\nabla\,(2)} = \frac{D^{\text{Old}}}{Dt}[\boldsymbol{\tau}]$).

Typically, Eulerian formulations of constitutive relations are implemented in FE systems by employing the updated Lagrangian approach (see, e.g., [7]) and using the fourth-order tangent stiffness tensors in two alternative forms of constitutive relations (see, e.g., [3, 6, 8])[1]

$$\frac{D^{\text{BH}}}{Dt}[\boldsymbol{\sigma}] = \mathbb{C}^{\text{BH}} : \mathbf{d}, \qquad \frac{D^{\text{Tr}}}{Dt}[\boldsymbol{\sigma}] = \mathbb{C}^{\text{Tr}} : \mathbf{d}. \tag{7.6}$$

Based on relations (2.26) between the stress rates, the tangent stiffness tensor for mixed models can be obtained from (7.4):

$$\mathbb{C}^{\text{Tr}}_{\text{mixed}} = \frac{2}{J}(\mu - \lambda \ln J)\mathbf{I} \overset{\text{sym}}{\otimes} \mathbf{I} + \lambda \chi(J)\mathbf{I} \otimes \mathbf{I}, \tag{7.7}$$

and that for vol-iso models can be obtained from (7.5):

$$\mathbb{C}^{\text{Tr}}_{\text{vol-iso}} = [K\,\chi(J) + \frac{2}{9}\mu \operatorname{tr}\mathbf{c}J^{-5/3}]\mathbf{I} \otimes \mathbf{I} + [\frac{2}{3}\mu \operatorname{tr}\mathbf{c}\,J^{-5/3} - 2K\,h'(J)]\mathbf{I} \overset{\text{sym}}{\otimes} \mathbf{I} \tag{7.8}$$

$$- \frac{2}{3}\mu\,J^{-5/3}(\mathbf{c} \otimes \mathbf{I} + \mathbf{I} \otimes \mathbf{c}).$$

Similar expressions for the tensors $\mathbb{C}^{\text{BH}}$ can be derived from expressions (7.7) and (7.8) using expressions (2.26) and (7.3)

$$\mathbb{C}^{\text{BH}} = \mathbb{C}^{\text{Tr}} + \mathbf{I} \overset{\text{sym}}{\otimes} \boldsymbol{\sigma} + \boldsymbol{\sigma} \overset{\text{sym}}{\otimes} \mathbf{I}.$$

Note that all the fourth-order tensors considered here have full symmetry.

Note the simplicity of expression (7.7) compared to expression (7.8). Note also that for mixed model #1,

$$\mathbb{C}^{\text{Tr}}_{\text{mixed}} = \frac{1}{J}\widehat{\mathbb{C}}^{(2)}_{\sharp},$$

where the tangent stiffness tensor $\widehat{\mathbb{C}}^{(2)}_{\sharp}$ is defined in Eq. $(68)_2$ in [6].

---

[1] In particular, the commercial Abaqus system uses rate constitutive relations of the form $(7.6)_1$ [9, 10], and the commercial MSC. Marc system uses rate constitutive relations of the form $(7.6)_2$ (cf., [11]). Due to (2.26), the positive definiteness of $\mathbb{C}^{\text{BH}}$ is equivalent to the Hill stability condition (see Chap. 5), so Abaqus uses the positive definiteness of $\mathbb{C}^{\text{BH}}$ as material stability condition.

7.2 General Conclusions

The main purpose of this study was to answer the question: are there compelling reasons to use the more complex neo-Hookean vol-iso model of compressible isotropic hyperelastic material rather than the simpler mixed model of the same material when simulating deformations of both rubber-like (slightly compressible) and foam-like (highly compressible) materials?

To answer this question, we performed a systematic study of the performance of both compressible neo-Hookean models, mixed and vol-iso, using seven well-known volumetric functions and a new one. The results of the study lead to the following conclusions.

First, in applications for simulating deformations of slightly compressible materials, both kinematic and static quantities obtained using the two types of models are close to each other as well as to the same quantities obtained using the incompressible neo-Hookean model.

Second, both types of models satisfy Hill's postulate in the same range of the volume ratio, and both types of models do not satisfy the corotational stability postulate (CSP) for all volumetric functions used in this study. Further, both model variants satisfy the polyconvexity requirement, provided that the volumetric term $h(J)$ is convex in J.

Third, compared to vol-iso models, mixed models have physically reasonable responses in extreme states for a wider set of volumetric functions. In particular, the popular volumetric function of the form $(\ln J)^2/2$ (not convex in J!) leads to physically reasonable responses for kinematic and static variables in extreme states when using mixed models, but does not lead to the same responses when using vol-iso models.

However, it should be noted that vol-iso models can be used in the range of Poisson's ratio $-1 < \nu < 0.5$, whereas mixed models can be used only in the range $0 \leq \nu < 0.5$.

To recapitulate, both the present work and previous study [12–14] have shown that when using volumetric functions from the Hartmann–Neff family with parameter $q \geq 2$ (the preferred value is $q = 5$ [15]), mixed and vol-iso models show similar performance in applications and have physically reasonable responses in extreme states, which is convenient for theoretical studies. However, mixed models allow the use of a wider set of volumetric functions with physically reasonable responses in extreme states, compared to vol-iso models. A second important advantage of mixed models over vol-iso models are more simple expressions for stresses and tangent stiffness tensors.

Note that the neo-Hookean material model has a narrow range application for simulating deformations of elastomers. First, its application is limited to simulating engineering strains of only about 10%, and, second, this material model does not take into account second-order effects, in particular the Pointing effect in simple shear or torsion of circular cross-section rods (see, e.g., [16]). The purpose of this study was to assess the feasibility of implementing different approaches to the simulation of deformations of compressible and slightly compressible elastomers in commercial FE systems. In particular, in MSC.Marc FE simulations of deformations using the generalized Mooney–Rivlin or Ogden models, compressible vol-iso material models are used for slightly compressible (rubber-like) elastomers and compressible mixed material models for compressible (foam-like) materials. Since the neo-Hooken model is a special case of both the Mooney–Rivlin and Ogden models, the present study shows that it is inappropriate to use models with different types of accounting for compressibility in FE systems. In fact, the simpler (in the mathematical sense) mixed Mooney–Rivlin or Ogden material models can equally successfully simulate deformations of both sufficiently compressible and slightly compressible materials.

References

1. Fong, J.T., Penn, R.W.: Construction of a strain-energy function for an isotropic elastic material. Trans. Soc. Rheol. **19**(1), 99–113 (1975). https://doi.org/10.1122/1.549389
2. Ogden, R.W.: Volume changes associated with the deformation of rubber-like solids. J. Mech. Phys. Solids **24**(6), 323–338 (1976). https://doi.org/10.1016/0022-5096(76)90007-7
3. Korobeynikov, S.N.: Analysis of Hooke-like isotropic hypoelasticity models in view of applications in FE formulations. Arch. Appl. Mech. **90**(2), 313–338 (2020). https://doi.org/10.1007/s00419-019-01611-3
4. Federico, S., Holthausen, S., Husemann, N.J., Neff, P.: Major symmetry of the induced tangent stiffness tensor for the Zaremba-Jaumann rate and Kirchhoff stress in hyperelasticity: Two different approaches. Math. Mech. Solids (2025). https://doi.org/10.1177/10812865241306703. (in press)
5. Simo, J.C., Pister, K.S.: Remarks on rate constitutive equations for finite deformation problems: computational implications. Comput. Methods Appl. Mech. Eng. **46**(2), 201–215 (1984). https://doi.org/10.1016/0045-7825(84)90062-8
6. Korobeynikov, S.N.: Families of Hooke-like isotropic hyperelastic material models and their rate formulations. Arch. Appl. Mech. **93**(10), 3863–3893 (2023). https://doi.org/10.1007/s00419-023-02466-5
7. Bathe, K.J.: Finite Element Procedures. Prentice Hall, New Jersey, Upper Saddle River (1996)
8. Korobeynikov, S.N.: Nonlinear Strain Analysis of Solids. Sib. Div. Russ. Acad. Sci, Novosibirsk (2000). (in Russian)
9. Ji, W., Waas, A.M., Bažant, Z.P.: On the importance of work-conjugacy and objective stress rates in finite deformation incremental finite element analysis. J. Appl. Mech. **80**(4), 041024 (2013). https://doi.org/10.1115/1.4007828
10. Nguyen, N., Waas, A.M.: Nonlinear, finite deformation, finite element analysis. Zeitschrift fü Angewandte Mathematik und Physik **67**(3), 35 (2016). https://doi.org/10.1007/s00033-016-0623-5
11. MSC.Software Corporation, Newport Beach (CA): MARC Users Guide. Vol. A. Theory and Users Information (2015)
12. Ehlers, W., Eipper, G.: The simple tension problem at large volumetric strains computed from finite hyperelastic material laws. Acta Mech. **130**(1), 17–27 (1998). https://doi.org/10.1007/BF01187040
13. Kossa, A., Valentine, M.T., McMeeking, R.M.: Analysis of the compressible, isotropic, neo-Hookean hyperelastic model. Meccanica **58**(1), 217–232 (2023). https://doi.org/10.1007/s11012-022-01633-2

14. Pence, T.J., Gou, K.: On compressible versions of the incompressible neo-Hookean material. Math. Mech. Solids **20**(2), 157–182 (2015). https://doi.org/10.1177/1081286514544258
15. Hartmann, S., Neff, P.: Polyconvexity of generalized polynomial-type hyperelastic strain energy functions for near-incompressibility. Int. J. Solids Struct. **40**(11), 2767–2791 (2003). https://doi.org/10.1016/S0020-7683(03)00086-6
16. Korobeynikov, S.N., Larichkin, A.Y., Rotanova, T.A.: Simulating cylinder torsion using Hill's linear isotropic hyperelastic material models. Mech. Time-Dependent Mater. **28**(2), 563–593 (2024). https://doi.org/10.1007/s11043-023-09592-1

Index

B
Bridgman's experimental data, 37

C
Cauchy–Green deformation tensor
   left, 1, 3, 14
   right, 1, 3
Configuration
   current, 14
   reference, 14
Continuous material spin tensors, 42

D
Deformation
   dilatational, 15
   isochoric, 15
Dilatation, 29
Double contraction, 9
Drucker's material stability condition, 41

E
Eigenindex, 10
Eigenprojections, 10
Eigenvectors, 10
Elastic energy, 1

G
Gradient
   deformation, 14
   spatial velocity, 15
Green elasticity, 1

H
Hill's stability condition, 33
Homogeneous deformation
   equibiaxial loading in plane stress (ELP), 55, 67
   uniaxial loading (UL), 55, 59
   uniaxial loading in plane strain (ULP), 55, 77
Hyperelasticity, 1
Hyperelastic material model, 3
   mixed, 4
   neo-Hookean, 3
   Ogden, 3
   vol-iso, 4

I
Inequality
   corotational stability postulate (CSP), 42
   Drucker's, 41
   Hill's, 41

L

Lamé parameter
  first, 18
  second, 19
Left stretch tensor, 14
Legendre–Hadamard ellipticity, 43

M

Material
  compressible, 18, 19
  incompressible, 18
  rate, 15
  time derivative, 15
Material model
  hyperelastic compressible
    Ciarlet–Geymonat's, 37
    Simo–Pister's, 37, 94
  incompressible Ogden, 23
  neo-Hookean
    compressible mixed, 25
    compressible vol-iso, 26
    incompressible, 23
Modulus
  bulk, 19
  shear, 18

P

Poisson's ratio, 28
Polyconvexity, 42, 43
Postulate
  corotational stability, 42
  corotational stability postulate (CSP), 5
  Drucker's, 5, 41
  Hill's, 5, 41
  TSTS-M$^+$, 42
Principal
  Cauchy stresses, 24
  stresses, 25, 26
  stretches, 14
Product
  double inner, 9
  external
    alternate, 10
    direct, 10
    dyadic, 10
  symmetric, 10
  inner, 9

R

Rank-one convexity, 33, 42
Rate constitutive relations
  for mixed models, 94
  for vol-iso models, 94
Richter–Flory decomposition, 2, 15

S

Spatial velocity vector, 15
Spectral representation, 10
Strain rate tensor, 16
Strain tensor
  Finger, 14
    deviator, 15, 24
    modified, 15
  infinitesimal, 16
  logarithmic (Hencky), 42
Stress tensor
  Cauchy, 16
  first Piola–Kirchhoff (1st PK), 16
  Kirchhoff, 16
  rate
    upper Oldroyd, 17
    Zaremba–Jaumann, 17
  weighted, 16
Stretching tensor, 16

T

Tensors
  coaxial, 11
  fourth-order
    fully symmetric, 9
    supersymmetric, 9
    tangent stiffness, 95
  orthogonal, 11
  second order
    skew-symmetric, 9
    symmetric, 9
  spherical, 15
  unimodular, 15

Index

U
Updated Lagrangian approach, 95

V
Volume ratio, 15
Volumetric functions family
    Hartmann–Neff's, 34
    Ogden's, 35
Vorticity tensor, 16

Y
Young's modulus, 28

MIX
Papier aus verantwortungsvollen Quellen
Paper from responsible sources
FSC® C105338

If you have any concerns about our products,
you can contact us on
ProductSafety@springernature.com

In case Publisher is established outside the EU,
the EU authorized representative is:
**Springer Nature Customer Service Center GmbH
Europaplatz 3, 69115 Heidelberg, Germany**

Printed by Libri Plureos GmbH
in Hamburg, Germany